AF683146

Electric Vehicle Design and Development

SP-862

All SAE papers, standards, and selected books are abstracted and indexed in the SAE Global Mobility Database.

Published by:
Society of Automotive Engineers, Inc.
400 Commonwealth Drive
Warrendale, PA 15096-0001
February 1991

ISBN 1-56091-126-3
SAE/SP-91/862
Library of Congress Catalog Card Number: 90-64299

Printed in USA

PREFACE

As the world acknowledges the economic impact of the diminishing availability of petroleum at affordable levels, it must also confront the ever increasing problems of noise pollution, air pollution and related atmospheric effects.

A recent report indicates that over 82 major U.S. urban areas are unable to meet established pollution standards. Current concerns about the green-house effect and emissions can be reduced by turning to electric vehicle technology. Improvements in the performance and range, as well as reduction in the cost, will help accelerate market acceptance of electric vehicles. Advancing the technology of the electric propulsion system is the key to improving these important vehicle characteristics.

The purpose of *Electric Vehicle Design and Development* (SP-862) is to aid further developments, stimulate thinking, and encourage future presentations. The committee wishes to extend a word of thanks to the authors and all others who have contributed to this publication and invite written submissions from any authors who have contributions to make to this body of knowledge. Such submissions will be considered for future SAE presentations.

Randolph D. MacDowall
in cooperation with
the SAE Electric Vehicle Committee

TABLE OF CONTENTS

910242

Performance Testing of the Vehma G Van Electric Vehicle

Gerald D. Whitehead and A. Scott Keller
Electrotek Concepts, Inc.
Chattanooga, TN

ABSTRACT

This paper presents the results of performance characterization testing of two prototype Vehma G Van electric vehicles (EVs). Testing was performed at the Electrotek Electric Vehicle Test Facility (EVTF) as part of the Electric Power Research Institute (EPRI)/Electrotek EV Program. The G Van is a GMC full-sized van converted to electric propulsion by Vehma International of Toronto, Ontario. One of the vans tested at the EVTF was a five-passenger model, the other was a two-seat cargo van. These vehicles utilize tubular-plate lead-acid batteries and dc powertrain system components produced by Chloride EV Systems of Redditch, England. Performance testing was conducted according to the EPRI/Electrotek EV Test Plan and included measurement of driving range at 56 km/h, on the SAE J227a C cycle, and on the Electrotek-defined Urban route. Vehicle top speed, maximum acceleration, dc and ac energy consumption, hill climbing, and braking capabilities also were measured. A series of product-improvement tests then was conducted. These tests compared driving range and energy consumption between the original-equipment passenger van and the same van that was modified by changing tire type, tire pressure, and rear axle type. Additional tests were performed to determine the power consumption and driving range effects of the vehicle's power steering/power brake unit and the air conditioner.

INTRODUCTION

Performance testing of electric vehicles (EVs) is one of the major work efforts of the Electrotek Concepts, Inc. (Electrotek) Electric Vehicle Test Facility (EVTF) in Chattanooga, Tennessee. The purpose of this testing is twofold: first, to monitor the advancing state of the art of EV technology; and second, to provide an independent source of vehicle performance data that will help prospective EV users make informed decisions about buying and operating EVs (1)*. This testing began in 1981 as part of the EV program started by the Tennessee Valley Authority (TVA) and the Electric Power Research Institute (EPRI). In 1988, Electrotek acquired the EV program staff from TVA and began operation of the EVTF through a licensing agreement with TVA. The EVTF is continuing to serve as EPRI's national EV and battery test center for the electric utility industry. In developing the EPRI/Electrotek EV Test Plan, Revision 2 (2), existing test procedures were combined with experience gained from previous tests to create a comprehensive evaluation plan. The resulting plan allows test results to be compared with data from other sources. Since the emphasis of this project is on real-world vehicle performance, testing is carried out primarily on the one-mile oval track located at the EVTF and on urban roads. Tests include constant-speed 56 km/h, SAE J227a C cycle, and real-world urban route driving range. Vehicle top speed, maximum acceleration, ac and dc energy consumption, gradeability (10 and 18 percent grades), and braking capabilities also are measured. In addition to the physical parameters being monitored (distance, speed, and time), electrical quantities such as power, energy (dc and ac), and charge (amperehours) are determined.

For the G Van evaluation, several tests were performed in addition to those described above. First, a series of product-improvement tests was conducted. These tests compared driving range and energy consumption between the original-equipment prototype passenger van and the same van that was modified by changing tire type, tire pressure, and rear axle type. Tests were then conducted to determine the power and energy consumption and driving range effects of the vehicle's power steering/power brake (PS/PB) unit and the air conditioner (A/C).

---

* Numbers in parentheses designate references at the end of the paper.

DESCRIPTION OF THE G VAN

The G Van is an electric delivery-type vehicle produced by Vehma International of Toronto, Ontario. This vehicle, which is in full compliance with all U.S. Federal Motor Vehicle Safety Standards (FMVSS), is based on a production GMC full-sized passenger van. Vehma's design philosophy was to maximize the use of standard GMC components and to minimize production complications in the installation of the electric propulsion system. To achieve this goal, special subframes were designed that are attached to the existing base-vehicle mounting points. All major electric-propulsion components are mounted to these subframes, which greatly facilitates vehicle assembly (3).

Both the passenger G Van, shown in Figure 1, and the cargo G Van have an integral construction chassis frame. Basic vehicle physical specifications are provided in Table 1. The rear-wheel-drive G Van utilizes the standard Vandura independent front suspension with upper and lower control arms, coil springs, and shock absorbers and the standard rear suspension with a solid rear axle, leaf springs, and shock absorbers. Steering is recirculating ball gear and utilizes the standard GMC diesel-van PS/PB hydraulic pump. This pump is belt driven by a 1.2 kW, 216-V dc motor. The standard Vandura front disc, rear drum mechanical brake system is used in the G Van.

The passenger van differs from the cargo van in the number of seats, the higher level of interior appointments, the additional side windows, and the corresponding increase in curb weight. Both vans have Eberspächer diesel-fueled heaters to provide interior heating. A 31-liter fuel tank is located under the van behind the rear axle.

Figure 1: Passenger G Van

Table 1
G Van Physical Specifications

	Original	Modified	Cargo
Gross Vehicle Wt. (kg)	3901	3901	3901
Curb Weight (kg)	3460	3451	3226
Max. Payload Wt. (kg)	441	450	675
Length (m)	5.1	5.1	5.1
Width (m)	2.0	2.0	2.0
Height (m)	2.1	2.1	2.1
Wheelbase (m)	3.2	3.2	3.2
Cargo Volume (m^3)	7.4	7.4	7.4

The G Van powertrain, shown in Figure 2, utilizes a Nelco separately-excited, 216-V dc motor rated at 45 kW. This motor is mated to a single-ratio gear-driven transmission which has a speed-reduction ratio of 1.85:1. The motor/transmission assembly is mounted beneath the cargo area floor at the right rear of the vehicle. A short drive shaft connects the transmission to the rear axle, which was rotated 180 degrees around the axle axis and installed with the drive shaft pointing to the rear of the vehicle.

A Chloride EV Systems (CEVS) Mk5A electronic motor controller is mounted under the hood of the vehicle in the original internal-combustion engine (ICE) location. The controller has transistor armature and field choppers. Vehicle direction is selected through the standard column-mounted shift lever (Park, Reverse, Neutral, and Drive) which reverses the polarity on the motor field when Reverse is selected. The controller protects the powertrain by limiting motor current at low speeds and in the event of excessive motor or controller temperatures. Regenerative braking is activated whenever the brake pedal is depressed while the vehicle is moving. Production G Vans are expected to be equipped with anti-lock rear brakes.

The battery pack, 36 six-volt CEVS 3ET205 lead-acid modules, is housed in a tray located beneath the cargo area floor. The tray is attached by four bolts (two in the front and two in the rear) to mounting brackets welded to the vehicle's floor pan and cross members. A unique energy-management system, which utilizes metallic shock absorbers and a front subframe energy absorber, was developed for the tray to provide high-speed impact attenuation while resisting the inertial loads produced in the vehicle during low-speed impacts (4). The vehicle must be raised from the ground and the battery weight supported from beneath for battery removal.

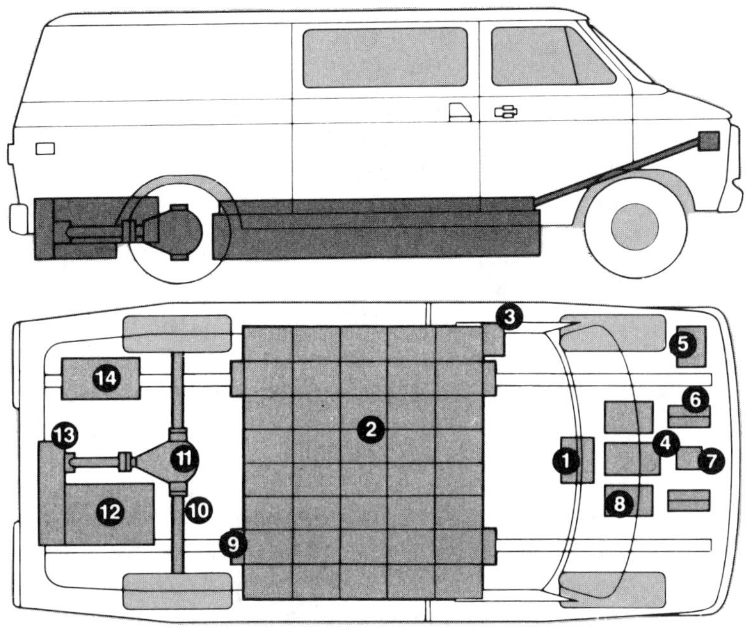

1 Traction Motor Controller
2 Detachable Battery Pack
3 Battery Pack Fuse
4 Power Steering/Power Brake Unit
5 12-V Auxiliary Battery
6 Battery Pack Watering/Venting System
7 Battery Pack Charging Plug
8 Diesel-Fueled Heater
9 Battery Attachment Point (One of Four)
10 Heavy-Duty Springs
11 Rear Axle
12 Electric Traction Motor
13 Gear-Driven Single-Speed Transmission
14 Heater Fuel Tank

Figure 2: G Van Component Layout

The 3ET205 battery is rated at 162.5 Ah at a 75-A discharge rate at 25°C. Electrical interconnections between modules are made through cables with cast-on lugs that are welded to the module terminals. The connectors and terminals are then encapsulated with a silicone rubber compound to provide protection against corrosion and improved handling safety. The 36 modules are separated into two independent watering paths and each string is subdivided into two five-module and two four-module strings by explosion arrest devices (EADs). These water traps and the dual-watering paths are designed to limit the possible propagation of a hydrogen explosion to a maximum of five of the 36 modules. The battery is watered by connecting one watering path at a time to a CEVS Autofil trolley via an under-hood coupler. The Autofil trolley, which includes a water reservoir and a vacuum pump, pulls water through the watering path to fill each cell. When all cells have been filled, the Autofil unit shuts off.

The traction battery is charged by an off-board 200/250-V, single-phase, 60-Hz Chloride Spegel S1P108/35 charger. The charger, which utilizes a taper-current charging profile, is connected to the battery through a socket located under the hood. For recharges from zero percent state of charge (SOC), nominal electrical characteristics are as shown in Table 2. The charger cutoff criteria is based on the rate of change of battery voltage (dv/dt). The typical recharge time from full discharge is nine hours. An override timer is provided to turn off the charger after 14 hours of operation if the dv/dt end-of-charge system fails. If the charger is left connected to the battery after completion of charge, it will compensate for inherent battery self discharge by turning on for 15 minutes every four hours. A safety interlock inhibits drive system operation while the charger is connected to the battery.

Table 2
Spegel Charger
Nominal Electrical Characteristics

Input Voltage	240 V ac
Initial Input Current	50 A ac
Initial Input Real Power	8.7 kW ac
Initial Output Current	35 A dc
Initial Output Power	7.8 kW dc
Final Input Current	20 A ac
Final Input Real Power	3.8 kW ac
Final Output Current	12 A dc
Final Output Power	3.3 kW dc

The vehicle controller has an integral dc-to-dc converter which charges the 12-V lead-acid auxiliary battery from the traction battery pack. Charging occurs both when the vehicle keyswitch is turned on and when the charger is plugged in.

TEST RESULTS

The two prototype G Vans were tested between September 1989 and November 1990 (5). During the evaluation period, the passenger van was driven a total of 9752 km and the cargo van was driven 4597 km. The passenger van was tested both as it was originally equipped from the factory and after several modifications were performed. These modifications, listed in Table 3, were intended to improve the vehicle's energy efficiency without the need for major structural or component alterations.

Table 3
G Van Product-Improvement Modifications

	Original	Modified
Tire Pressure (bar)	3.8 F/5.2 R	5.5 F/5.5 R
Tire Type	General 8.75R16.5L	Michelin XCH4 LT 225/75 R16
Rear Axle Type	GM Saginaw 4.56:1	BTR 4.714:1

The cargo van, which was loaned to the EPRI/Electrotek program by the Los Angeles Department of Water and Power, was equipped with General tires inflated to 5.5 bar front/rear and a BTR rear axle. The only change made during the testing was to install Michelin tires inflated to 5.5 bar front/rear for a comparison with the General tires.

Test results are summarized in Tables 4 and 5, given on the next page. Both vehicles were tested at curb weight plus one-half maximum payload. The data reported are averages of two repetitions for each test. Note that for lead-acid batteries, available battery capacity increases with temperature. It is therefore necessary to adjust battery discharge and vehicle range data to a standard temperature of 25°C. This adjustment allows these data to be compared without including the effects of battery temperature variations. The correction factor used for temperature adjustment is based on the end-of-test battery temperature; this factor is given below. Measured test results that are independent of battery temperature are shown in Table 4. Measured test results that are a function of battery temperature, adjusted to 25 °C, are given in Table 5. Although ambient temperature has an effect on the energy consumption of the vehicle, its effect over the relatively narrow range of test temperatures is minimal compared with the effects of battery temperature. Therefore, no provision is made for adjusting the test results based on ambient temperature. Due to running changes in the test plan and time/weather constraints, not all tests were conducted for each vehicle.

$$\text{Correction Factor} = 1/(1+(0.00792 \times (T_F - 25)))$$

where T_F = final battery temperature (°C)

The driving range tests were terminated when the battery voltage dropped below a predefined level at a particular current, as described in the Appendix. According to the SAE J227a test procedures, the C cycle test, shown in Figure 3, also could be terminated if the vehicle could not meet the prescribed acceleration.

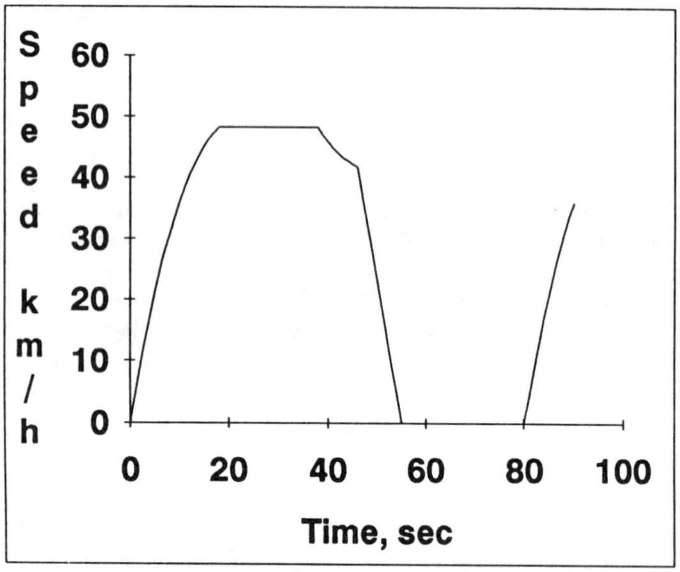

Figure 3: SAE J227a C Cycle

The 10-percent-grade test determined how far the modified G Van could travel on a 10-percent grade at a constant 24 km/h. This test involved driving the van on a chassis dynamometer to simulate an infinitely-long 10-percent grade. Based on actual road-test data, the dynamometer was loaded to 34 kW for this evaluation.

Figure 4 gives the constant-speed dc power characteristics for the three different G Van configurations. Figure 5 shows the constant-speed dc energy consumption, a measure of vehicle energy efficiency, for the three vehicle configurations. Figure 6 shows the energy consumption of the vehicles on one cycle of the SAE J227a C Cycle. As expected, the original-equipment G Van required the most power at each test speed and it had the highest energy consumption.

Table 4
G Van Test Results, Temperature-Independent Quantities

	Original	Modified	Cargo
Curb Weight (kg)	3460	3451	3226
Test Weight (kg)	3682	3682	3530*
Top Speed, 100% SOC (km/h)	86.0	84.5	88.7
Top Speed, 20% SOC (km/h)	86.3	84.8	89.5
Top Speed for 5 Minutes, 100% SOC (km/h)	-	84.0	88.4
Top Speed for 5 Minutes, 20% SOC (km/h)	-	84.0	87.4
Acceleration Time 0 to 48 km/h, 100% SOC (seconds)	12.2	12.1	11.9
Acceleration Time 0 to 48 km/h, 20% SOC (seconds)	12.4	12.2	12.1
Top Speed on 18% Grade, 100% SOC (km/h)	10.9	-	-
Top Speed on 18% Grade, 20% SOC (km/h)	10.8	-	-
Braking Distance from 48 km/h (m)	16.7	-	-
56 km/h ac Energy Consumption (Wh/km)	470	390	416
56 km/h dc Energy Consumption (Wh/km)	271	232	234
C Cycle ac Energy Consumption (Wh/km)	693	650	576
C Cycle dc Energy Consumption** (Wh/km)	366	329	335
Urban ac Energy Consumption (Wh/km)	987	952	920
Urban dc Energy Consumption** (Wh/km)	518	400***	520
Measured Battery Capacity at 75 A, Adjusted to 25 °C (Ah)	168.2	168.5	169.6

*This prototype cargo van was tested at the planned production-version curb weight, plus half payload, of 3530 kg rather than its actual half-payload weight of 3564 kg.

**These are net quantities calculated using discharge energy minus regenerative energy.

***Instrumentation problems caused this apparent erroneous result.

Table 5
G Van Test Results, Temperature-Dependent Quantities

	Original	Modified	Cargo
Constant-Speed 56 km/h Range (km)	130.6	144.6	158.7
SAE J227a C Cycle Range (km)	84.5	91.1	97.6
Urban Route Range (km)	54.3	49.7*	59.1
10-Percent Grade Range (km)	-	8.4	-

*The motor overheated and began leaking oil during these tests.

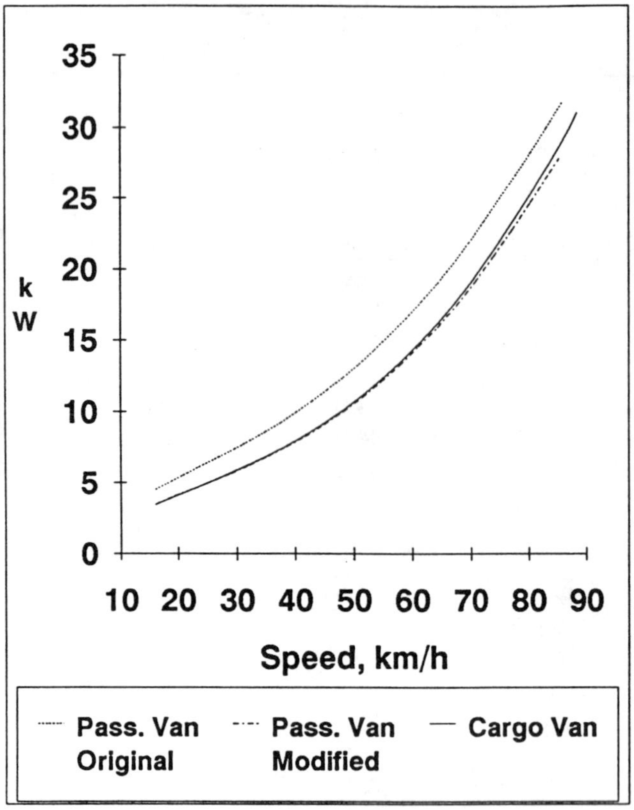

Figure 4: Constant-Speed dc Power Characteristics

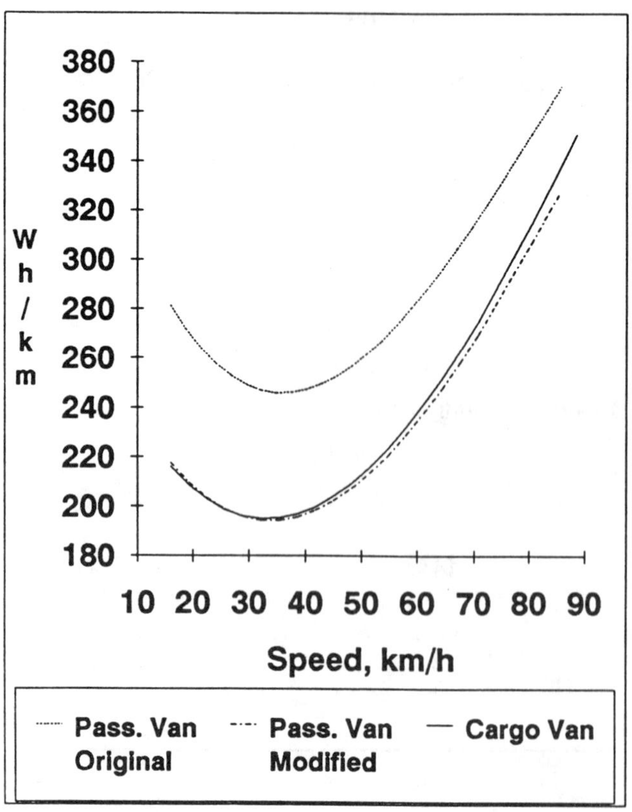

Figure 5: Constant-Speed dc Energy Consumption

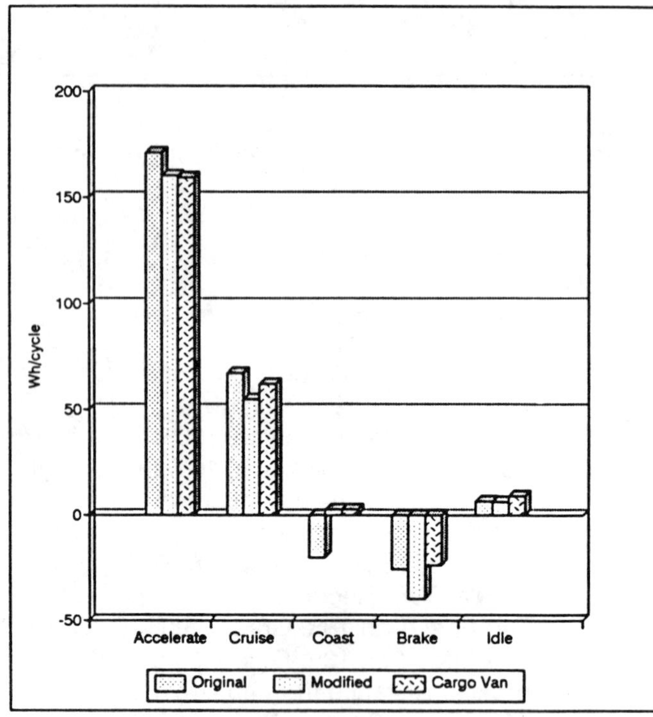

Figure 6: C Cycle Energy Consumption

As can be seen from Figure 6, there was significant energy usage during the idle portion of the C cycle. For example, the modified van consumed 1.2 kWh during idle over a full test. (Also evident from Figure 6 is that the original van had regenerative braking during vehicle coasting. This feature was disabled for compatibility with the planned anti-lock braking system.) A series of tests was performed to identify which components contributed to the energy usage during idle, with the results being presented in Figures 7 through 9. Figure 7 shows a breakdown of the power consumption for all 12-V auxiliaries. Figure 8 is a comparison between the power consumption of the auxiliaries and the idle power of the PS/PB. Figure 9 is a comparison between the power consumption of the auxiliaries and full power requirements of the PS/PB. Note that the power requirements for the PS/PB were the same for any of the following conditions: wheels straight, full brakes; wheels full turn, no brakes; or wheels full turn, full brakes.

Because of the high idle-power requirements for the PS/PB unit, a series of C cycle range tests was conducted with the modified G Van to determine its effect on driving range. Significant results from these tests, which were run both with the PS/PB unit operating and with it disabled, are given in Table 6. These data are adjusted to a battery temperature of 25 °C.

A similar series of tests was conducted with the modified G Van to determine the effects of the A/C on C cycle range; temperature-adjusted test results are shown in Table 7. Additionally, the A/C's cool-down capabilities were determined by "soaking" the van in direct sunlight for three hours prior to each test. The ambient temperature ranged from 30 °C at the start of these tests to 38 °C at the end; the relative humidity averaged 55 percent. From an initial interior (driver's area) temperature of 46 °C, the A/C lowered the temperature to 25 °C within 13 minutes. The driver's area temperature was maintained at 18 to 21 °C for the remainder of test.

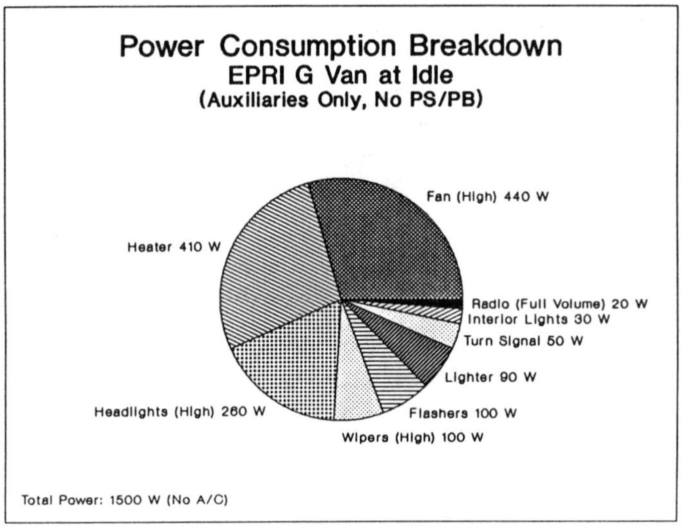

Figure 7: Auxiliary Component Power Consumption

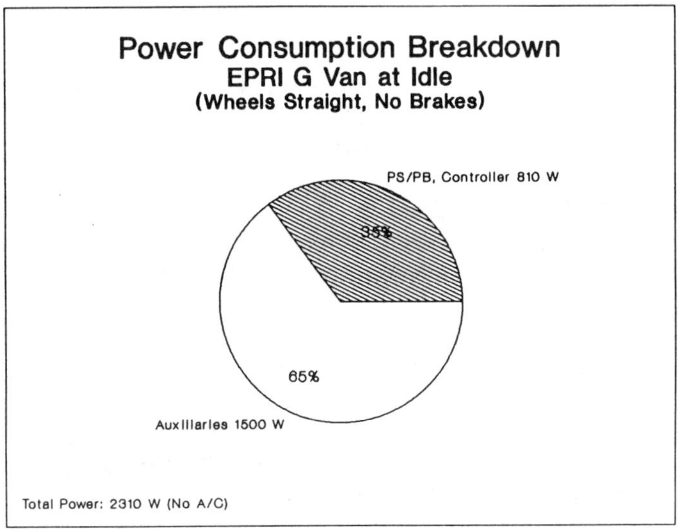

Figure 8: Power Consumption of Auxiliaries vs. PS/PB (Idle)

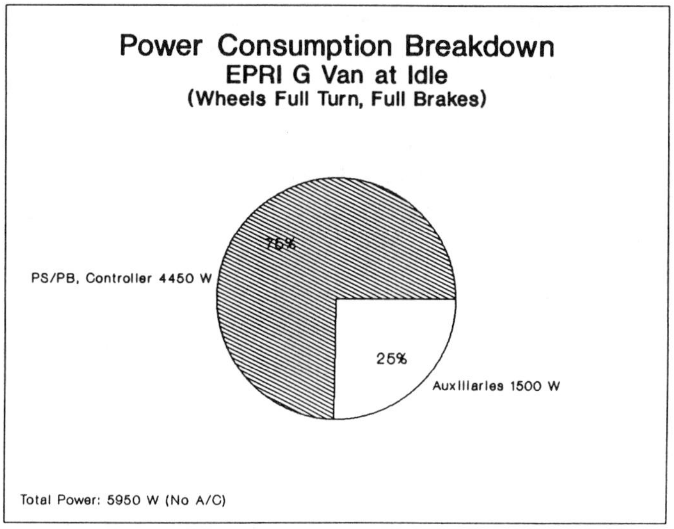

Figure 9: Power Consumption of Auxiliaries vs. PS/PB (Full Turn, Full Brake)

Table 6
Power Steering/Power Brake
C Cycle Range Results

	PS/PB On	PS/PB Off
Driving Range (km)	87.9	99.0
dc kWh Discharge	36.2	37.6
dc kWh Regeneration	7.1	9.2
dc Energy Consumption (Wh/km, net)	329	286

Table 7
A/C C Cycle Range Test Results

	A/C On	A/C Off
Driving Range (km)	70.8	94.5
dc kWh Discharge	32.6	36.9
dc kWh Regeneration	3.6	5.9
dc Energy Consumption (Wh/km, net)	410	328

As testing of the G Vans progressed, the constant-speed energy consumption rates of the vehicles were observed to decrease. This was due to the gradual break-in of the various components, resulting in fewer electrical and mechanical losses. As can be seen in Figure 10, there was a 13.0 percent reduction in energy consumption at 32 km/h for the cargo van between tests run at 2582 km and at 4492 km.

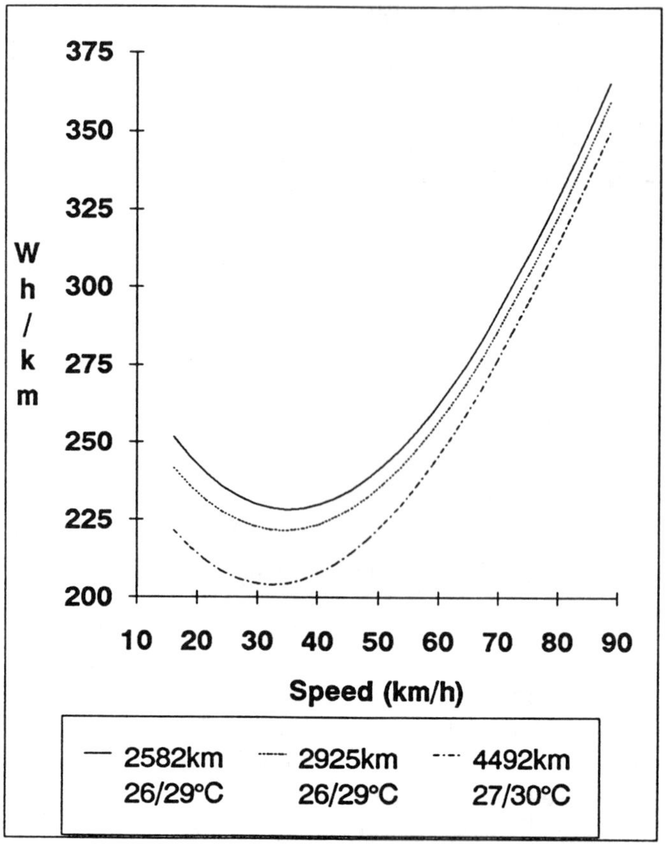

Figure 10: Effects of Break-In on Energy Consumption

CONCLUSIONS

Two Vehma G Vans, a five-passenger model and a two-seat cargo van, were performance tested at the Electrotek EVTF as part of the EPRI/Electrotek EV Program. These vehicles, which utilize tubular-plate lead-acid batteries and dc powertrains, have suitable performance to serve in various urban "niche-market" fleets as delivery or service vans. The vans are well engineered, as evidenced by their successful completion of the FMVSS and Industrial Standards/Durability Tests. The top speed, acceleration, and braking capabilities of these vehicles allow them to be safely integrated into urban traffic.

A successful product-improvement program was conducted with the original prototype passenger van to enhance its performance. These changes, which increased the vehicle's C cycle driving range by 7.8 percent (84.5 to 91.1 km, temperature adjusted to 25 °C), will be implemented into production-version vans.

Additional tests were performed to determine the effects of the PS/PB unit and the A/C on driving range and energy consumption. It was found that the PS/PB system decreased C cycle driving range by 11.2 percent. Use of the A/C was determined to reduce C cycle range by 25.1 percent when the average ambient temperature was 34 °C.

Plans now call for a limited production run of G Vans beginning in early 1991. If these initial vehicles perform successfully, they will pave the way for future EVs to replace ICE vehicles in many fleet applications, thus helping to ease ever-increasing environmental concerns.

ACKNOWLEDGMENTS

The authors wish to acknowledge the contributions during the G Van evaluation of J. Hampton Barnett, Director of the EVTF; Theodore W. Blickwedel and Harshad S. Tataria, EVTF Battery Test Engineers; Willie R. Idlette, Jr. and Gary H. Barnett, technicians; and Katherine H. Aracri, EVTF secretary.

REFERENCES

1. R.L. Driggans, G.D. Whitehead, "Performance Characterization Testing of the GM Griffon and G Van," November 1988, EVS-9 Proceedings, Toronto, Ontario.

2. G.D. Whitehead, J.H. Barnett, "Electrotek/EPRI EV Project Electric Vehicle Test Plan (Revision 2)," November 1989, Electrotek, Chattanooga, TN.

3. W. Harney, "Electric G Van Development for Production, Phase 1 Summary Report," November 1988, Magna Universe, Toronto, Ontario.

4. S. Palvoelgyi, P. Stangl, "Crashworthiness of Electric Vehicles," April 1990, SAE Truck Engineering.

5. G.D. Whitehead, A.S. Keller, "Test and Evaluation Report for the Vehma G Van Electric Vehicle," February 1991, EPRI, Palo Alto, CA.

APPENDIX

Driving range and battery capacity tests require a standard method for deciding when the battery has reached 0 percent SOC. The end-of-test criteria utilized by Electrotek involves monitoring the voltage at a particular current. For tubular-plate lead-acid batteries, the nominal cutoff point at a 75-A discharge rate is 1.65 V per cell. For the G Van, this corresponds to a pack voltage of 178.2 V. Cutoff voltages for different discharge currents are listed in Table A-1.

Table A-1
G Van End-of-Test Voltage/Current Criteria

Current	Voltage
10	185.2
20	184.1
30	183.1
40	182.0
50	180.9
60	179.8
70	178.7
80	177.7
90	176.6
100	175.5
110	174.4
120	173.3
130	172.3
140	171.2
150	170.1
160	169.0
170	167.9
180	166.9
190	165.8
200	164.7

910243

Track and Dynamometer Testing of the Eaton DSEP Minivan and Comparisons with Other Electric Minivans

A. F. Burke and R. D. MacDowall
Idaho National Engineering Laboratory
EG&G Idaho, Inc.
Idaho Falls, ID

ABSTRACT

Track and dynamometer testing of the Eaton Dual Shaft Electric Propulsion (DSEP) minivan has been performed by the Idaho National Engineering Laboratory (INEL). The dynamometer testing included constant speed tests up to 88 km/h and driving cycle tests on the SAE J227a C and D cycles and the Federal Urban Driving Schedule (FUDS) as well as maximum effort acceleration tests. The dynamometer data were analyzed to determine the energy consumption (Wh/km) of the DSEP vehicle for the various driving modes and to project the range of the vehicle if the NiF170 nickel-iron battery had been at its rated capacity. Ranges of 90-125 miles at constant speeds and about 70 miles on the driving cycles were projected.

Comparisons were made of the performance of the DSEP vehicle and the ETX-II and the TEVan minivans, which have been developed on other DOE and EPRI programs using lead-acid, nickel-iron, nickel-cadmium, and sodium-sulfur batteries. The comparisons utilized test data when available, but were based primarily on computer simulation results obtained using the SIMPLEV program.

INTRODUCTION

Research and Development programs leading to proof-of-concept electric propulsion systems for battery powered minivan vehicles have been funded by the United States Department of Energy (DOE) and the Electric Power Research Institute (EPRI) since 1984. The vehicles associated with these programs are the DSEP (Dual Shaft Electric Propulsion) built by the Eaton Corp.,the ETX-II built by Ford Motor Co. and General Electric, and the TEVan built by Chrysler Pentastar. The first of these programs to be completed was the Eaton DSEP in September 1989. This paper is concerned with the testing of the DSEP by the Idaho National Engineering Laboratory. Tests of the DSEP were performed on the track at the Chrysler Proving Grounds in Phoenix, Arizona and on the chassis dynamometer in Idaho Falls. The data obtained in those tests are analyzed in detail and compared with computer simulations of the operation of the DSEP under various conditions.

The ETX-II and TEVan vehicles have not yet been tested to the same extent as the DSEP, but some data and computer projections are available for those vehicles for comparison with the data for the DSEP. Those comparisons are made in this paper for several different battery types in the vehicles.

THE EATON DSEP ELECTRIC MINIVAN

Vehicle and Propulsion System Description

The DSEP electric vehicle was built from a Chrysler T-115 minivan. It utilizes a three phase alternating current (AC) induction motor, a transistorized, microprocessor controlled inverter, and a two-speed, automatically shifted transaxle. The nickel-iron (NiF) battery pack, which is positioned under the floor of cargo area, consists of 28 6V modules having a rated cell capacity of 170 Ah. The floor of the production minivan was raised slightly to accommodate the battery pack resulting in a minimal loss of vehicle utility. The exterior appearance of the DSEP is unchanged from the T-115 van. Hence the DSEP has the utility of a ground-up design even though it is essentially a conversion from a production vehicle. The vehicle and powertrain characteristics of the DSEP are summarized in

Table 1. Three vehicles (see Figure 1) were built by Eaton in the DSEP program including the TB-1 which served as a mule for the first generation propulsion system. The complete DSEP development program is covered in detail in the final report, Reference 1.

Computer Simulations

Computer simulations of the operation of the DSEP have been run at INEL using the PC program SIMPLEV (Reference 2). Propulsion system component inputs for the calculations are based on inverter-motor and transaxle efficiency data (Reference 3) taken in the laboratory by Eaton for various motor torques and speeds. The vehicle data inputs (weight, C_DA, and tire rolling resistance) are based on coastdown data taken by INEL. SIMPLEV simulates vehicle operation second-by-second for various driving profiles-constant speed and the SAE J227a and FUDS driving cycles. The results of these calculations are used later in the paper for comparison with the test data.

The battery characteristics used in the paper are based on tests of NiF-170 modules at Argonne National Laboratory (ANL) as well as at the INEL. By the time the DSEP was delivered to INEL for testing, the capacity of the NiF-170 battery pack had degraded to less than 100 Ah making it necessary to use data taken at ANL to characterize the pack in its rated condition. Battery characterization tests of the pack as used in the INEL vehicle tests were made in the INEL Battery Test Laboratory (Reference 4), but the data were used primarily to assess the condition of the pack rather than its like-new performance, which was needed for the simulation inputs (see Table 2).

TRACK TESTING

Facilities and Instrumentation

The TB-2 DSEP vehicle was transported in November 1989 to the Chrysler Proving Grounds in Phoenix, Arizona for coastdown and constant speed track testing. The track facilities used in Phoenix are shown in Figure 2. The coastdown tests were performed on a 2.4 mile straight, essentially level, section of the track. Multiple tests were run in both directions of the track to allow averaging of any grade and wind effects. The constant speed tests were done on a 1.6 mile oval track having $\pm$ 1% grades on the short straight sections and both grade and bank on the closed end sections (see Figure 2).

The vehicle was instrumented with a fifth-wheel (Nucleus Corp. Model NC8) to measure velocity. The coastdown data were collected on a Campbell Scientific Data Logger (Model 21X). Weather information (ambient temperature, atmospheric pressure, and wind speed and direction) were obtained from an on-site Campbell Scientific weather station. For the constant speed tests, the vehicle was powered by the NiF170 nickel-iron battery pack. During the tests, battery voltage, current, and power were recorded in addition to vehicle velocity. Basically the same battery instrumentation that was used in the dynamometer tests in Idaho Falls was used for the constant speed track tests in Phoenix. All data were sampled at the 10 Hz rate, but average values for 1 second intervals were stored in the Campbell Scientific Data Logger for later transfer to a PC computer for storage and analysis.

Coastdown Tests

The coastdown tests were performed to determine the rolling resistance and aerodynamic drag coefficient of the DSEP vehicle. Tests were performed at three weights (2909, 2620, and 1795 kg) as a means of determining the magnitude of the velocity dependent term in the rolling resistance equation for the tires (see Appendix I for a discussion of the rolling resistance results). The coastdown data were analyzed using the methods given in Reference 5. It was determined from the data that the coastdown curve and road load of the DSEP vehicle could be calculated to sufficient accuracy (with an error of less than two percent) using a C_DA of 12 ft2 and a velocity independent tire rolling resistance coefficient of .0095 for speeds up to 60 mph (96 kmph). These values were used in all the simulations and to calculate the road load (Figure 3) when needed. The actual coastdown curve (Figure 4) measured for the vehicle weight of 2620 kg (5763 lbs.) and tire pressure of 35 psi was used to setup the dynamometer for the laboratory tests of the DSEP.

Constant Speed Track Tests

Constant speed tests of DSEP on the 1.6 mile oval track in Phoenix were run at speeds of 48, 64, and 88 kmph for comparison with dynamometer tests. Measured battery power as a function of time during typical constant speed tests are shown in Figure 5. The large variations in power are thought to be due primarily to the grades on the straightaways and banked oval sections of the track and secondarily to changing wind speed and direction as the vehicle travels around the track. The period of the oscillating battery power curve matches that expected by the time required for the vehicle to travel to/from the two curved ends of the oval track. As indicated in Figure 5, the amplitude of the power oscillations are much larger than those experienced during constant speed tests on the dynamometer. The average energy consumption (Wh/km) on the track at essentially constant speed are 20-25% higher than was measured on the dynamometer at the same speed. This difference will be discussed further in a later section of the paper.

DYNAMOMETER TESTING

Facility, Instrumentation, and Test Setup

Dynamometer tests of the DSEP vehicle were performed in the INEL Vehicle Dynamometer Laboratory in Idaho Falls. A photograph of the laboratory with the DSEP on the dynamometer is shown in Figure 6. The dynamometer is an electromechanical Clayton chassis dynamometer with road load of form-$A+B*V+C*V^2$ - provided by a computer-controlled DC motor. The A, B, and C parameters, which are input from the keyboard used to control the dynamometer, are varied during the dynamometer calibration procedure until the coastdown curve of the vehicle on the dynamometer matches that measured on the track.

The primary quantities measured in the DSEP dynamometer tests were battery voltage, current, and power and dynamometer roll torque and RPM. The electrical quantities are obtained using a special power measurement instrument (PMI) developed especially for electric vehicle testing (Reference 6). Battery voltage and current signals are input into the PMI, which processes them and outputs digital or analog signals proportional to battery voltage, current, and power. The current sensor is either a coaxial shunt or a Hall-effect transducer (Lem Model LT-200). Measurements of the dynamometer torque and RPM are needed by the dynamometer control system so those signals are readily available for use in the vehicle tests. The dynamometer laboratory data acquisition system (DAS) is built around a DEC LSI-11/73 minicomputer. It utilizes 16-bit architecture which supports the RSX-11M operating system. The DAS is configured for monitoring 128 measurement channels. The analog-to-digital converter has a resolution of 11 bits plus sign (0.05% FS conversion accuracy). Low (20 mv) and high (10 V) level channels are multiplexed through the A/D converter. Data are stored on a 30-Mbyte Winchester Disk Drive from which it can be displayed on the terminal console or output after the test to a printer. Data are archived on a nine-track tape drive system. The DAS and instrumentation used for vehicle testing at INEL are discussed in detail in References 7 and 8.

Considerable care is required in setting up the dynamometer for electric vehicle tests and then confirming that the setup yields the road load appropriate for the vehicle being tested. In principle, all that is required is that the dynamometer be set at the proper inertia weight (2620 kg in the case of the DSEP) and that the values of A, B, and C be determined for which the coastdown of the vehicle on the dynamometer with the half-axles removed matches that measured on the track. This requires a number of test runs on the dynamometer at various values of A, B, and C. Since the effective vehicle road forces on the dynamometer are the sum of those due to the tires on the rolls and the DC motor, care must be taken to assure that the tire temperature is the same for all the dynamometer coastdown runs and that the tire temperature selected is close to the average expected for the vehicle dynamometer tests to be done later. The force of the tires on the rolls at a given vehicle speed can be determined by measuring the dynamometer parasitic losses with and without the vehicle (with half-axles removed) on the rolls. The total road load of the vehicle on the dynamometer is then obtained by adding the tire force to that due to the DC motor, which includes the dynamometer spin losses. This has been done for the DSEP and the resulting road load compared to that calculated from the track coastdown data. As shown in Figure 7, there is excellent agreement between the road load on the track and dynamometer. Hence it was concluded that the dynamometer setup was satisfactory.

Constant Speed and Energy Economy Tests

The constant speed and energy economy tests of the DSEP vehicle on the dynamometer were intended to determine the energy consumption (Wh/km) and powertrain efficiency at various speeds. That information can be used to project the range of the vehicle using different batteries and compare the efficiency of different electric propulsion systems at various power levels.

The constant speed tests were made at 48 km/h, 64 km/h, and 88 km/h. The energy economy tests were run at speeds between 32 km/h and 88 km/h with selected speeds being held constant for 60-120 seconds after stabilization; the speed steps are repeated after the battery has been discharged to lower states-of-charge. As will be discussed below, the energy consumption at a particular speed changes significantly as the powertrain and tires warm-up. The potential effect of warm-up on the data should be considered in comparing vehicle energy consumption data from various sources.

As shown in Figure 8, the variation in battery power during a constant speed test on the dynamometer is quite small-seldom being more than a few tenths of a kW from the mean value. This is in marked contrast to the large variations in battery power experienced in the constant speed track tests (see Figure 5). The constant speed tests on the dynamometer approach the ideal from the point-of-view of constant power demand, but they are less than ideal in terms of propulsion system and tire warm-up. As shown in Figure 9, the vehicle energy consumption decreases with time during the test. The higher initial values are due primarily to the higher rolling resistance of the cold tires, which heat up during the test (see Figure 10). The energy consumption value reported for the test is either the average for the test or the value at a time midway through the test before the tire temperatures greatly exceeded the temperature used in the dynamometer coastdown calibration.

Energy economy data for DSEP are shown in Figure 11 for two test periods-one near the beginning of the test and a second about 4500 seconds into the test after the tires are warmed-up. Also shown on the figure are the average energy consumption values for the constant speed tests at 48, 64, 88 km/h. These latter values agree well with the energy economy data taken in the second time period. The energy consumption values for DSEP are summarized in Table 3. Also shown in the table are the values of Wh/km calculated using the SIMPLEV simulation program along with the corresponding powertrain efficiencies determined using the road load power given in Figure 3. The efficiency values vary from about 55% at light loads to 80% at the higher loads. This indicates that the DSEP propulsion system-battery to the wheels-is quite efficient. As shown in Figure 12, the efficiencies inferred from the dynamometer data agree reasonably well with those given by the SIMPLEV simulation.

The energy consumption for the the DSEP vehicle determined from the constant speed track tests in Phoenix are given in Figure 13 compared with the values obtained in the constant speed dynamometer tests. The track values are 20-25% higher than the dynamometer values. This is an uncommonly large difference based on past experience. It is felt that the reason for this difference is the character of the track (see Figure 2) on which the tests were performed. The track had 1% grades on the straight sections and banked curves with variable grade. This resulted in large variations of battery power as the vehicle traveled around the track at constant speed (see Figure 5) in contrast to the dynamometer tests in which the both the speed and power were essentially constant (see Figure 8). The INEL dynamometer and track test and SIMPLEV simulation results are compared in Figure 14 with the energy consumption values presented by Eaton in Reference 1 based on testing at the Chrysler Proving Grounds in Chelsea, Michigan. The large scatter (as indicated by the scatter bars) in the Eaton data make it of little value in assessing the validity of the INEL dynamometer constant speed energy consumption data. Hence since both the INEL and Eaton track data for the DSEP appear flawed, the dynamometer data, supported by the SIMPLEV simulations, are at the present time the best information available on the energy consumption of the DSEP vehicle at various constant speeds.

Driving Cycle Tests

Dynamometer tests of the DSEP vehicle were run using the Federal Urban Driving Schedule (FUDS) and the SAE J227a C and D cycles. Energy consumption data for those driving cycles are given in Figure 15 and Table 4. As in the case of the constant speed tests, the energy consumption decreases with time as the tires and drive system warm-up. Driving cycle test results are affected significantly by the power required when the vehicle is stopped (V=0 or idle periods in the driving cycle) and the fraction of energy returned to the battery by regenerative braking. The idle power requirement of the DSEP is about 2.3 kW, which is quite high. This is due to the desire to maintain a positive motor torque when the vehicle is sitting at a stop. If this feature in the software were eliminated, the idle power could be reduced to about 300 W and the energy consumption improved. The improvement varies with the driving cycle as shown in Table 4, but is significant. The fraction of energy recovered by regenerative braking varies between 4 and 6% for the driving cycles tested. This is a relatively low energy recovery as a potential for 10-15% recovery exists and has been achieved in some vehicles (References 9-10).

Vehicle Range

The range data obtained in the dynamometer tests of the DSEP are not a good indication of the vehicle's range potential, because the NiF170 batteries in the vehicle at the time of the tests had degraded to a capacity of about 90 Ah from their rated capacity of 170 Ah. The energy capacity of the battery pack was only about 15 kWh instead of the expected 30 kWh. The measured range and the projected range of the DSEP based on the measured energy consumption values and a NiF170 battery having rated performance (see Figure 16) are given in Table 5. The measured range was only about half the potential range of the vehicle. As shown in Table 5, a small improvement in range on urban driving cycles would result from reducing the idle power requirement to 300 W from the present value of 2.3 kW.

Maximum Effort Acceleration Tests

Maximum effort acceleration tests were run on the dynamometer. The DSEP vehicle was accelerated to 98 km/h (61 mph). The measured acceleration times are summarized below:

0-32 km/h	3.1 (3.2)	sec
0-48 km/h	6.5 (7.5)	sec
0-64 km/h	12.7 (11.5)	sec
0-80 km/h	22.9 (18.0)	sec
0-88 km/h	28.6 (n.a.)	sec

The acceleration times measured on the track by Eaton (Reference 1) are shown in ( ) in the table above. The reason for the differences in the acceleration times at the higher speeds is that in the INEL tests the battery could sustain the peak power of 65 kW for only about 8 seconds, while in the Eaton tests on the track the battery sustained 65 kW for the entire 18 seconds of the test. This is another indication of the degraded condition of the battery for the INEL tests. SIMPLEV simulations of the DSEP yielded acceleration times of 8 sec for 0-48 km/h and 19.5 sec for 0-80 km/h with a peak battery power of 62 kW. The maximum motor output power was 55.5 kW at 7000 RPM in the simulations.

COMPARISON OF THE DSEP WITH OTHER ELECTRIC MINIVANS

Vehicle Characteristics

The DSEP, ETX-II, and the TEVan were built starting with ICE-powered minivans. The DSEP and TEVan are based on the Chrysler Caravan/Voyager and the ETX-II is based on the Ford Aerostar. The characteristics of the three electric minivans and their propulsion systems are given in Table 6. The chassis weights, including the propulsion system, but excluding the batteries, are based on published weight breakdown information (References 1,11-14) on the vehicles. The ETX-II is the lightest of the vehicles by about 160 kg (300 lbs) primarily, because the base Ford Aerostar is lighter than the Chrysler Caravan/Voyager by approximately 300 lbs. The peak motor power capability of the three vehicles is essentially the same being about 70 Hp (52.5kW). The acceleration characteristics of the vehicles, especially at speeds greater than 80 km/h, are dependent as much on the shape of the maximum torque and power curves at high motor RPM as on the peak motor power. Unfortunately, test data on the operation of the drivelines at maximum power at high motor RPM are not available for all the driveline systems. A good source of such data is maximum effort acceleration vehicle testing on the dynamometer.

Vehicle Simulations

The comparisons of the performance of the three minivans will be made based on available test data and computer simulations. If sufficient test data were available, the comparisons could rest primarily on data with the simulations used to include the effect of advanced batteries on vehicle performance. At the present time, only limited test data are available for the ETX-II and the TEVan, so the performance characteristics given in this paper for those vehicles are based primarily on simulation results. When possible, the simulation results are validated with the available test data on the three minivans. All the simulations presented have been run using the program SIMPLEV, which is described in some detail in Reference 2. The input data for the simulations are based on published data on the vehicles and their driveline components, including the batteries, as well as unpublished data available at INEL obtained directly from the vehicle developers under contract to DOE and EPRI and other national laboratories working on electric vehicles and batteries. The data bases for the DSEP and ETX-II are quite extensive and complete. In the case of the TEVan, the data base for the motor, electronics, and especially the transmission is much less complete and therefore, there is more uncertainty regarding the validity of the TEVan simulation results than for the DSEP and ETX-II. The battery modeling is based on test data taken at INEL and Argonne National Laboratory, so it is felt the calculated battery performance is realistic for "new" batteries having rated capacity. No attempt has been made to include battery degradation in the simulations.

Performance Comparisons

The performance comparisons for the three minivans are made on the following basis:

1. Chassis weight, including the driveline, is vehicle specific reflecting the best information for that vehicle. A payload of 272 kg (600 lbs) is used for all the runs.

2. The aerodynamic parameter, C_DA, is vehicle specific based either on INEL or manufacturer coastdown data. The

ambient density used in the calculations corresponded to standard atmospheric conditions (.07657 lb/ft3).

3. The same tires were used on all the vehicles. The rolling resistance of the tires was taken as .0095 lb/lb based on the coastdown data taken by INEL on the DSEP vehicle. The rolling resistance was taken to be speed independent. (see Appendix A for a discussion of this aspect of tire rolling resistance)

4. For each type of battery, the same technology is used in all the vehicles. If volume and system voltage constraints permit, the same battery pack (cell Ah capacity , number of modules, and battery weight) is used. When a variation in battery pack between vehicles is necessary, a minimum change is used.

Simulations were made for the following battery technologies:

1. Flooded Lead-Acid (Japan Storage ED 150P, 12V)
2. Nickel-Iron (Eagle-Picher, NiF200)
3. Nickel-Cadmium (Saft STM 5-200, semi-sealed)
4. Sodium-Sulfur (CSPL 10 Ah cell, 300 Ah bank)

The characteristics of these batteries are summarized in Table 7 based on the information and data given in References 15-21. These batteries represent technologies that are either commercially available or are being seriously considered for commercialization by battery developers around the world. From the performance point-of-view, these batteries all have sufficiently high power density (peak W/kg) for electric vehicle applications, but they vary greatly in energy density (Wh/kg). The batteries also vary considerably in cost, cycle life, and environmental impact, but those factors were not considered in the present study, which was concerned primarily with vehicle performance.

Energy Consumption

A good measure of the performance of an electric vehicle is the energy consumption (Wh/km) of the vehicle for different types of driving. Energy consumption for the three minivans are compared in Table 8 for constant speeds of 56, 72, and 88 km/h and the SAE J227a C and FUDS driving cycles. The same Sodium-Sulfur battery pack (625 kg, CSPL-ETX-II) was used in all the simulation runs, so the differences in energy consumption reflect vehicle specific factors such as chassis weight and propulsion system component efficiencies. The simulation results indicate that the TEVan has the lowest energy consumption for all modes of driving. There is not sufficient test data available for the TEVan to confirm this low energy consumption with confidence. The data given in Reference 14 are track data for a vehicle test weight of 2727 kg (6000 lbs). That data indicated an energy consumption of 165-173 Wh/km at 56 km/h and 255-295 Wh/km on the C cycle. The corresponding simulation values for a vehicle weight of 6000 lbs are 160 and 245 Wh/km. Both the data and the simulation results for 6000 lbs indicate the TEVan has a lower energy consumption than the DSEP would have at the same weight, but the magnitude of the difference is uncertain and must await more test data on the TEVan and its driveline components, especially the transmission. A comparison between the simulation results and Ford test data for the ETX-II (Reference 12) is given in Table 9. The differences between the data and the predictions are relatively small being less than 10% in nearly all instances. The simulation results and data available for the electric minivans indicate their energy consumption is quite predictable and that the differences between the various designs will not be large.

Range

The ranges of the minivans using various types of batteries are shown in Table 10 based on the simulation results. The volume of each battery pack is about 400 liters with a pack voltage of about 200 V. The cell capacity (Ah) was selected to be consistent with the voltage and volume constraints. The vehicle weights varied to reflect the differences in the weights of the different battery packs. The range values are shown for constant speeds of 56 and 88 km/h and the FUDS driving cycle. The ranges vary significantly with battery type, but not as much as might have been expected based on the energy density of the different battery types. This occurs because the higher energy density batteries have a lower density (kg/liter) resulting in a lower battery weight for the 400 liter battery volume. The differences in range for the lead-acid, nickel-iron, and nickel-cadmium battery packs for the same driving cycle are not large especially on the FUDS. The vehicles using sodium-sulfur batteries have ranges that are 60-70% greater than those using lead-acid batteries and 40-50% greater than those using nickel-cadmium or nickel-iron batteries. If rapid charging of the ambient temperature batteries in about one hour is considered an option, the range advantage of the sodium-sulfur battery is greatly reduced and electric vehicles with daily maximum ranges of 150-200 miles become possible using the ambient temperature batteries.

Acceleration Times

The acceleration characteristics of the three minivans are summarized in Table 11. The acceleration times depend primarily on vehicle weight and powertrain peak power with the effect of battery type reflected only through its influence on vehicle weight. All the batteries studied have sufficiently high power density that for current vehicle designs, the peak power of the battery pack does not limit vehicle acceleration. If the peak powertrain output powers were significantly increased from present values (50-55 kW), this would not be the case and battery power characteristics could limit vehicle acceleration performance. Acceleration characteristics are given in Table 11 for two vehicle weights spanning the values for designs using sodium-sulfur and lead-acid batteries. All three minivans have essentially the same acceleration performance - 0-48 km/h in 7-9 seconds, 0-80 km/h in 15-19 seconds, 0-96 km/h in 28-31 seconds. Comparisons of track test data and simulation results for the three vehicles are given in Table 12 . The simulations were run at the vehicle weights at which the track tests were performed. The agreement between the data and the predicted acceleration times is good for all the vehicles.

ACKNOWLEDGEMENT

Work supported by the U. S. Department of Energy under contract number DE-AC07-76ID01570

REFERENCES

[1.] Final Report- Advanced Dual Shaft Electric Propulsion System Technology Development Program, Contract DE-AC07-84NV10366, Eaton Corporation, September 1989

[2.] Burke, A.F. and Cole, G.H., Simulation of Electric Vehicles with Hybrid Power Systems, Proceedings of the 10th International Electric Vehicle Symposium, Hong Kong, December 3-5, 1990

[3.] DSEP TB-1 Powertrain Test Report, Eaton Corporation, May 1987

[4.] Hardin, J.E., Laboratory Tests of the NiF170 6V Traction Battery, EG&G Idaho Report No. EGG-EP-9371, January 1991

[5.] Venhuizen, J.R., Techniques to Analyze Vehicle Coastdown Data, EG&G Idaho Report No. EGG-ED-0725, April 1985

[6.] Griffin, D.C. and Bryant, J.A., Data Acquisition System for Electric Vehicle Tests, Proceedings of the IAS Annual Meeting, IEEE Industry Applications Society, September 1980

[7.] Crumley, R.L., INEL Electric Vehicle Laboratory Qualification: Transition Test Results and Comparisons with ETV1-2 Data, EG&G Idaho Report No. EGG-SE-6743, November 1984

[8.] Heiselmann, H.W., Acceptance, Verification, and Performance Test Results for the Bedford CF Electric Van, DOE Report No. DOE/ID-1041, January 1986

[9.] The DOE ETV-1 - Phase III Final Report (Performance Testing and System Analysis), Jet Propulsion Laboratory Report, DOE/CS54209-3, December 1981

[10.] Driggans, R. and Whitehead, G., Test and Evaluation Report for the GM G Van, TVA Report, June 1988

[11.] Single-Shaft Electric Propulsion System Technology Development Program - ETX-II, Phase II Technical Report, Contract DE-AC07-85NV10418, Ford Motor Company, April 1989

[12.] Single-Shaft Electric Propulsion System Technology Development Program -ETX-II, Final Technical Report, Contract DE-AC07-85NV10418, Ford Motor Company, October 1990

[13.] TEVan Program Review Materials, September 16, 1988, presented by the Pentastar Electronics Inc.

[14.] TEVan Program Review & Ride and Drive, April 4, 1990, Chelsea, Michigan, presented by Pentastar Electronics, Inc.

[15.] Brochure on Electric Vehicle Batteries, Japan Storage Battery Company, Kyoto, Japan

[16.] Cornu, J.P., The Ni/Cd Battery: The Actual Best Electrochemical Generator for EV, Proceedings of the 10th International Electric Vehicle Symposium, Hong Kong, December 3-5, 1990

[17.] Hardin, J.E., Laboratory Tests of the SAFT SEH-5-200 6 Volt Traction Battery, EG&G Idaho Report No. EGG-EP-8917, December 1989

[18.] Deluca, W.H. etals, Performance Evaluation of Advanced Battery Technologies for Electric Vehicle Applications, Proceedings of the 25th IECEC, Volume 3, Reno, Nevada, August 12-17, 1990

[19.] Mangan, M.F., Sodium Sulfur Batteries for Electric Road Vehicles, Proceedings of the 9th International Electric Vehicle Symposium, Toronto, Canada, November 13-16, 1988

[20.] ETX-II Design Review, Chloride Silent Power Limited, Washington, D.C., February 28, 1989

[21.] Deluca, W.H., Preliminary Test Results on CSPL 64-V Na/S EV Battery, presented at the CSPL Na/S Design Review Meeting, Washington, D.C., February 28, 1989

[22.] White, R. A. and Korst, H. H., The Determination of Vehicle Drag Contribution from Coastdown Tests, SAE Paper No. 72099, Automotive Engineering Conference, January 10-14, 1972, Detroit, Michigan

[23.] Korst, H. H. and White, R. A., Aerodynamic and Rolling Resistances of Vehicles as Obtained from Coastdown Experiments, Presented at the Second International Conference on Vehicle Mechanics, Paris, France, Sept. 1971

[24.] Road Load Measurements and Dynamometer Simulation Using Coastdown Techniques--SAE J1263, Report of the Passenger Car, Light Duty Truck Fuel Economy Measurement committee, June 1979

[25.] Menendes, P., Rolling Resistance Test Data for Michelin XA4 and XZX Tires, Private Communication to R. D. MacDowall from Michelin, June 8, 1990

APPENDIX A

DISCUSSION OF ROLLING RESISTANCE AND DRAG COEFFICIENT DETERMINATION FROM COASTDOWN TEST DATA

As is discussed in the literature (Reference 22-23), the determination of the rolling resistance and drag coefficients of a vehicle from coastdown data is not simple even when there are no problems with grade or wind on the track. The reason for this is that both the aerodynamic and tire rolling resistance forces have some velocity squared dependency. In the case of the rolling resistance force, the form of velocity dependency is not known precisely as it varies with tire construction and materials. In general, the rolling resistance coefficient fr varies as

$$fr=C0+C1*V+C2*V^2$$

and the vehicle road load force Frl as

$$Frl=W*fr+ 1/2\ \rho/g\ C_D*A\ V^2$$

One approach to determining the coeffcicents C0, C1, C2, and CD for a vehicle is to perform coastdown tests at several vehicle weights as a means of uncoupling the rolling resistance and aerodynamic forces. Coastdown tests of the DSEP vehicle were performed at weights of 3951, 5755, 6400 lbs by removing the battery pack from the vehicle and adding some additional pay load up to the GVW of the vehicle. The analysis of the coastdown data obtained is discussed in this Appendix.

The total road load force as a function of vehicle velocity can be determined (Reference 5) from the coastdown curves (velocity vs. time) for the different vehicle weights. The C0 cefficient can be determined by extrapolating the force curve to zero velocity and dividing the force obtained by the vehicle weight. This yields the velocity independent rolling resistance force which is then subtracted from the total road load force to obtain the velocity dependent forces for each coastdown. The effect of vehicle weight on the velocity dependent forces is then found by subtracting for each velocity, the velocity dependent forces at the various vehicle weights. The force differences, which are due to tire rolling resistance only as it is assumed the aerodynamic forces are independent of vehicle weight, are then divided by the vehicle weight differences to obtain the proportionality factor between the vehicle weight and the velocity dependent rolling resistance forces at each velocity. Using this proportionality factor, the velocity dependent rolling resistance force can be calculated and subtracted from the total velocity dependent road load force to obtain the aerodynamic force at each velocity. Knowing the air density of the tests, the effective C_DA for each speed and vehicle weight can then be calculated. It was found that the average C_DA for the set of coastdown curves was 11.65 with a scatter of only ± 0.15 for speeds of 40 mph and higher. The aerodynamic force can now be calculated for each speed and subtracted from the total velocity dependent force to obtain the velocity dependent rolling resistance force for each speed and vehicle weight. Dividing this force by the vehicle weight and plotting the results as a function of vehicle speed, one obtains Figure A-1, which indicates that the dominate velocity dependent portion of the rolling resistance varies as V^2. This finding is consistent with SAE J1263 for measuring road load using coastdown techniques (Reference 24), which recommends a rolling resistance expression of the form - $C1(1+C2*V^2)$ - and specifies a value of C2 unless it is determined separately by tire tests. Curvefitting the data points in Figure A-1 using linear regression analysis yields for the present tests of the DSEP vehicle, the following expression for the rolling resistance

$$fr = .0092 + .9*10^{-6}*V + 1.6*10^{-7}*V^2$$

This relation, combined with a C_DA of 11.65, yields coastdown curves in excellent agreement with the coastdown data (see Figure A-2).

To obtain additional data on the rolling resistance of commercially available high performance radial tires like those used on the DSEP vehicle, INEL requested data from Michelin for their XA4 and XZX series tires. Data (Reference 25) were obtained from Michelin for those tires tested on drum-roll and twin-roll dynamometers. The data are shown in Figure A-3 in terms of fr as a function of speed up to 80 mph. The tires on the DSEP vehicle as tested were equivalent to the XA4 series, which is classified as an all-season radial by Michelin. As shown in Figure A-3, drum-roll dynamometer data agree very well with results of the coastdown tests for the tires on the track. This would seem to validate the approach used to extract the coefficient data from the coastdown curves at the different vehicle weights. The Michelin data also show clearly that the rolling resistance of a tire on a twin-roll dynamometer is much higher than on a drum-roll dynamometer or the track. This is not an unexpected result.

The primary reasons for the coastdown tests are to obtain a coastdown curve for setting the chassis dynamometer and information from which the road load can be determined for simulation of the vehicle on the computer. The first of these objectives can be met more easily than the second, because the coastdown curve can be used directly if the coastdown test is performed at the test weight to be used in the dynamometer tests. Meeting the simulation objective is more difficult as it requires extracting the rolling resistance and aerodynamic coefficients from the coastdown data. In either case, what is really needed is the road load as a function of vehicle speed. In the present study, it was decided to simplify the coastdown results by using a C_DA of 12 ft2 and a velocity independent rolling resistance coefficient of 0.0095, which yielded a road load vs. speed curve in good agreement with the coastdown data (see Figure 3). This also permits the calculation of the coastdown curve for other vehicle weights and ambient densities.

Table 1. Characteristics of the DSEP Vehicle

Vehicle Test Weight	2620 kg (5763 lbs)
Battery	
NiFe - NiF-170	
28 modules	700 kg
Maximum power	65 kW
Drag Coefficient	.375
Frontal Area	$2.97m^2$ ($32\ ft^2$)
Rolling Resistance	.0095
Accessory Load	300 W
Idle Load	2.3 kW
Powertrain	
Motor	
Type - AC Induction	
Max. RPM - 12000	
Peak Power - 55 kW	
Inverter	
3-phase, variable voltage, variable frequency	
400 A peak rms current	
300 V max. voltage	
Transaxle	
Automatic, 2-speed	
Gear ratios: 24.83 in low, 13.99 in high	

Table 2. NiF-170 Inputs for SIMPLEV

NIF-170	
5	cells per module
24.1	module weight (kg)
170	Ah capacity
277	Bconst
-.11195	B Exp
	$Ah = Bconst\ (I_{av})^{BEXP}$
2	C rating

Depth of Discharge	VOC	Resistance (Ω)
0	1.36	.00051
.1	1.345	.00051
.2	1.329	.00051
.3	1.315	.00051
.4	1.300	.00050
.5	1.290	.00050
.6	1.280	.00050
.7	1.260	.00054
.8	1.240	.00058
.9	1.176	.00065
1.0	1.112	.00073

Table 3. Energy Economy and Powertrain Efficiency Values for the DSEP Vehicle (Dynamometer Tests)

Speed (km/h)	Energy Consumption (Wh/km) Test	SIMPLEV*	Powertrain Efficiency* (%)
32	156	149	53
40	157	149	59
48	157	152	65
56	160	165	72
64	168	186	76
72	180	196	81
80	197	217	83
88	214	235	83

*ambient density - .072 lb/ft^3

Table 4. Energy Consumption Values for the DSEP Vehicle on Various Driving Cycle from the Dynamometer Tests and SIMPLEV Simulations

Driving Cycle	Energy Consumption[1] (Wh/km) Dynamometer Tests	SIMPLEV[3]	% Energy Recovery in Tests
SAE J227a C	298 (270)[2]	255	3.9 (6.5)[4]
SAE J227a D	276 (267)	265	6.1 (8.2)
Federal Urban Driving Schedule (FUDS)	296 (285)	267	5.2 (7.1)

(1) Coastdown for ρ = .0721 lbm/ft^3

(2) Corrected to 300 W auxiliary load during idle

(3) Used 300 W auxiliary load

(4) Percent of energy recovered from regenerative braking - Test (SIMPLEV)

Table 5. Range of the DSEP Vehicle - Measured and Projected

Driving Mode	Range (km) Measured[1]	Projected[2]
Constant Speed		
48 km/hr	107	212
72 km/hr	94	179
88 km/hr	65	144
Driving Cycle		
SAE J227a C	53	112 (124)[3]
SAE J227a D	51	121 (125)
FUDS	50	112 (116)

(1) With NiF-170 batteries as received in the DSEP

(2) With NiF-170 batteries at rated capacity

(3) Range with 300 W idle auxiliary load

Table 6. Characteristics of the Electric Minivans

Characteristic	DSEP	ETX-II	TEVan
Vehicle Weight w/o batteries (kg)	1375	1227	1375
Payload (kg)	273	273	273
Drag Coefficient	.375	.370	.375
Frontal Area (m^2)	2.97	3.25	2.97
Tire Rolling resistance (%)	.95	.95	.95
Nominal Voltage	182	200	195
Auxiliary Load (W)	300	250	200
Max. Motor Power (kW)	55	52.5	52.5
Gear Ratios: 1st	24.4	21.2	15.2
2nd	14.0	13.8	8.4

Table 7. Characteristics of the Batteries Used in the Minivans

	BATTERY TYPES			
Characteristic	Lead-Acid	Nickel-Iron	Nickel-Cadmium	Sodium-Sulfur
Typical Battery	Japan Storage ED-150P	Eagle-Picher NiF-200	SAFT STM 5-200	CSPL ETX-II
Reference No.	15	18	16,17	19-21
Wh/kg	35	52	65	90
Wh/ℓ	77	94	100	135
(W/kg)max	150	100	200	100
(W/ℓ)max	330	180	320	150
kg/ℓ	2.2	1.8	1.6	1.5
Volume (liters)	400	382	422	422
Weight (kg)	875	685	685	625

Table 8. Energy Consumption of the Minivans for Various Driving Modes

	Energy Consumption (Wh/km)		
Driving Mode	DSEP* (2318 kg)	ETX-II (2125 kg)	TEVan (2318 kg)
Constant Speed			
56 km/h	162	173	141
72 km/h	190	221	175
88 km/h	235	270	222
Driving Cycles			
C-Cycle	235	248	209
FUDS	247	253	220

* All vehicles use sodium-sulfur batteries (575 kg)

Table 9. Comparison of the Energy Consumption of the ETX-II Vehicle from Dynamometer (Ford) Test Data and SIMPLEV

	Energy Consumption (Wh/km)	
Driving Mode	Ford Test Data	SIMPLEV
Constant Speed		
56 km/h	165	171
72 km/h	202	219
88 km/h	299	269
Driving Cycle		
C-Cycle	229	244
FUDS	247	249

Table 10. Range of the Minivans Using Different Batteries

	Range (km)								
	DSEP			ETX-II			TEVan		
Battery Type(1)	56(2)	88	FUDS	56	88	FUDS	56	88	FUDS
Lead-Acid (ED-150P)	204	117	138	146	74	106	238	128	152
Nickel-Iron (NiF-200)	246	155	164	---	---	---	285	169	179
Nickel-Cadmium (SAFT 5-200)	240	155	160	224	133	154	276	169	174
Sodium-Sulfur (CSPL-ETX-II)	342	225	211	321	195	210	396	245	235

(1) All battery packs approximately 400 liters

(2) Speeds in km/h

Table 11. Acceleration Times for the Minivans

			Acceleration Times (sec)			
Vehicle	Peak Motor Power (kW)	Vehicle Wgt (kg)	0 - 48(1)	0-80	0-88	0-96
DSEP	55	2318(2) 2500	7.2 7.9	17.5 19.2	22.2 24.3	29 32
ETX-II	52.5	2125 2375	5.7 6.4	15.3 17.2	19.4 21.19	25.9 28.8
TEVan	52.5	2318 2500	7.7 8.6	17.9 20.0	22.0 24.4	28 31

(1) All speeds in km/h

(2) Vehicle weights with sodium-sulfur and lead-acid batteries

Table 12. Comparisons of Acceleration Times from Test and SIMPLEV

		Acceleration Times (sec)			
		0-48 km/h		0-80 km/h	
Vehicle	Test Weight (kg)	Track	SIMPLEV	Track	SIMPLEV
DSEP	2495	7.5	8.2	20	19.9
ETX-II	2045	7.0	6.3	17.4	18.2
TEVan	2727	8.2	9.6	24	22.0

Figure 1. Photograph of the Three DSEP Vehicles

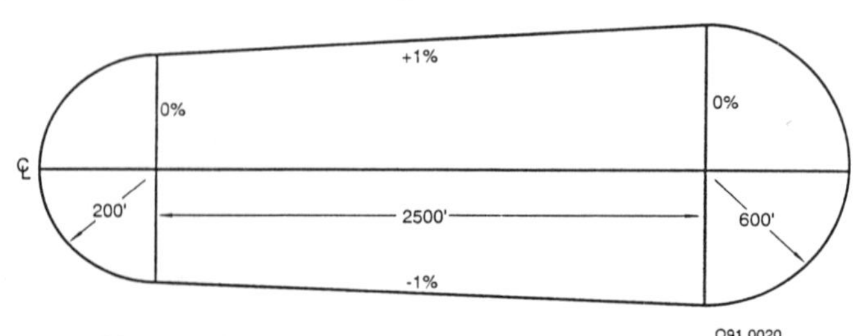

Figure 2. Layout of the Chrysler Proving Ground track in Phoenix, Arizona

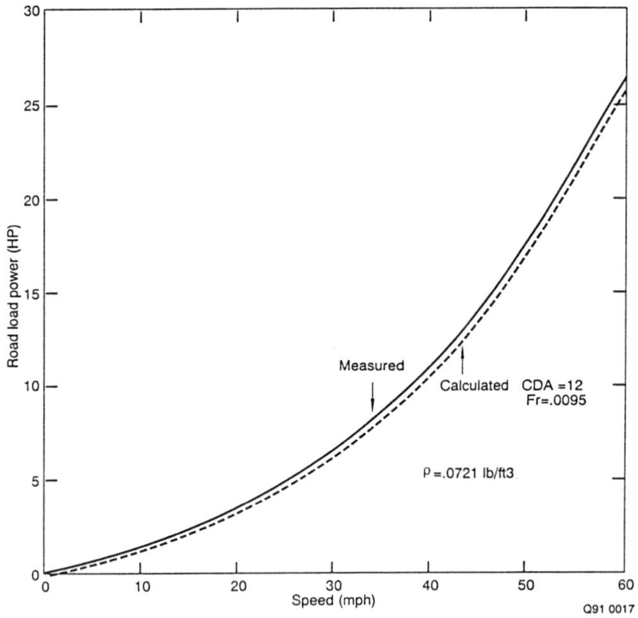

Figure 3. Measured and calculated road load power of the DSEP vehicle

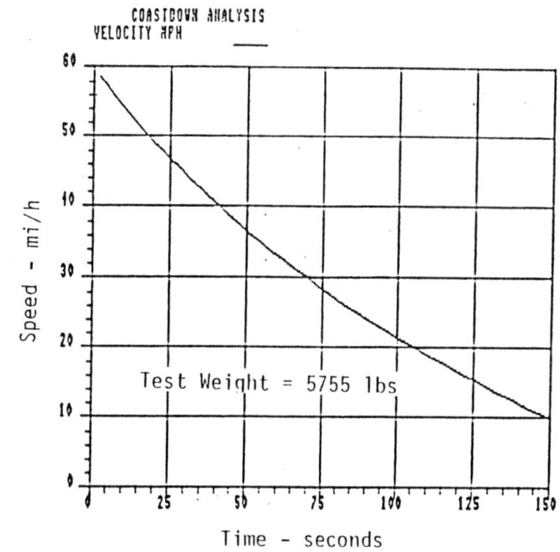

Figure 4. Coastdown Curve for the DSEP vehicle on the track

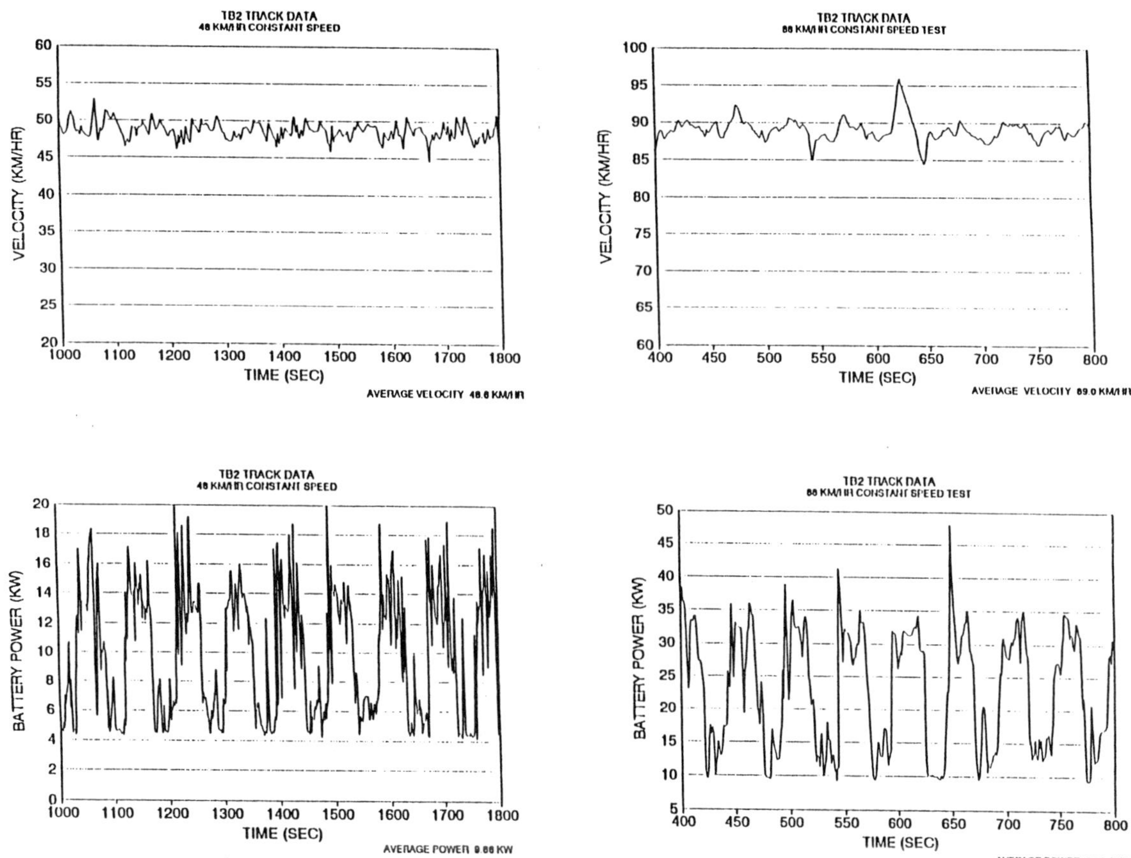

Figure 5. Speed and battery power variations during constant speed test on the track at Phoenix, Arizona

Figure 6. Photograph of the DSEP Vehicle on the Dynamometer

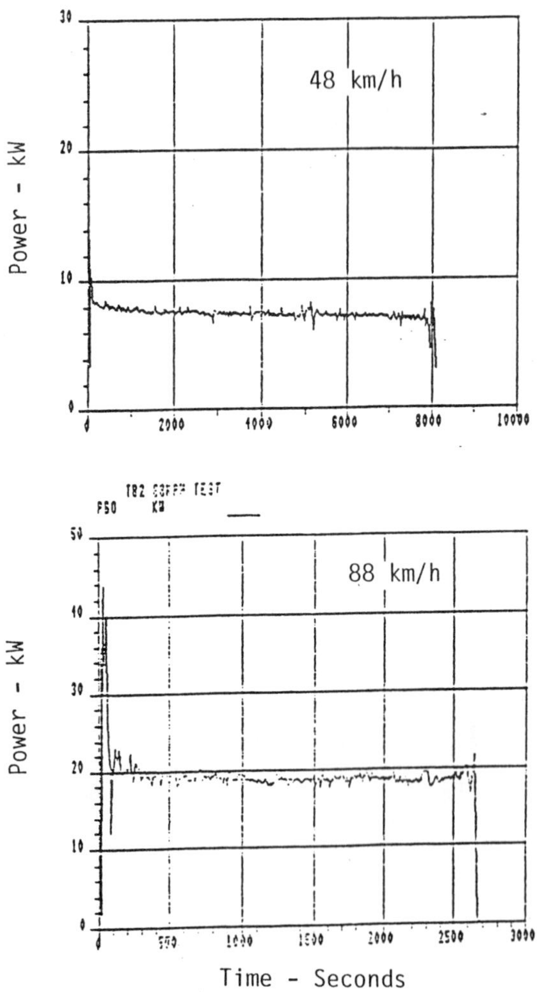

Figure 8. Battery power during a constant speed test on the Dynamometer

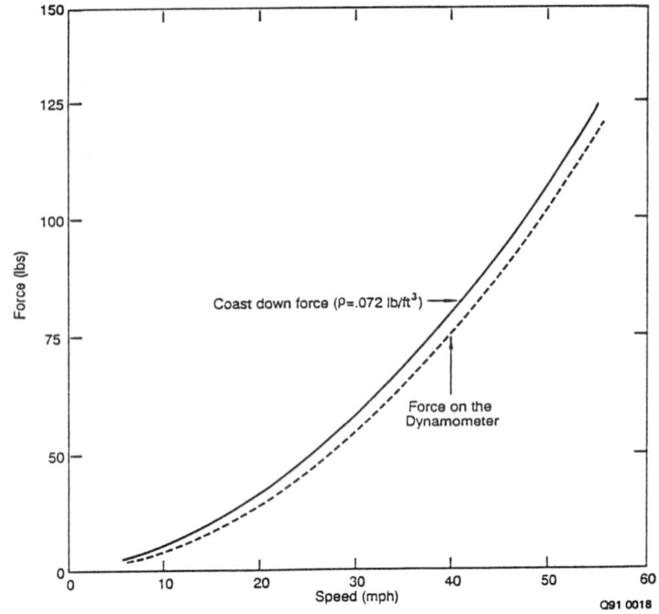

Figure 7. Comparison of the Dynamometer and coastdown road load forces

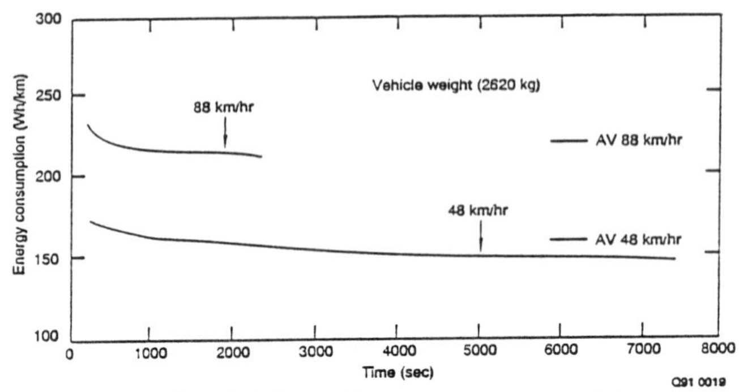

Figure 9. Variation of energy consumption during constant speed test on the Dynomometer

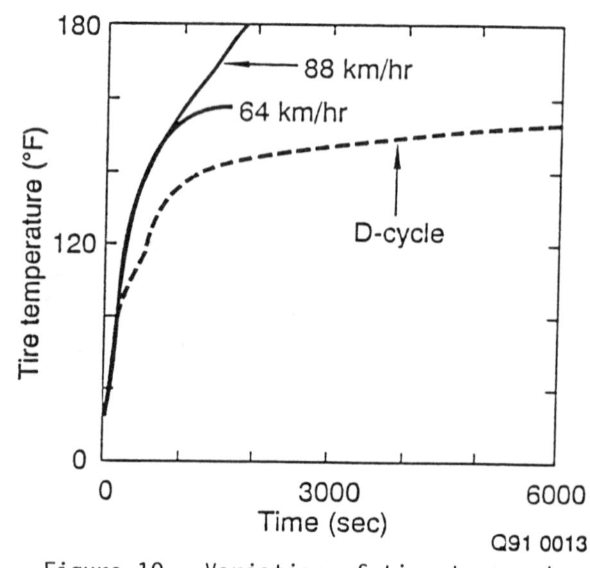

Figure 10. Variation of tire temperature during vehicle Dynamometer tests

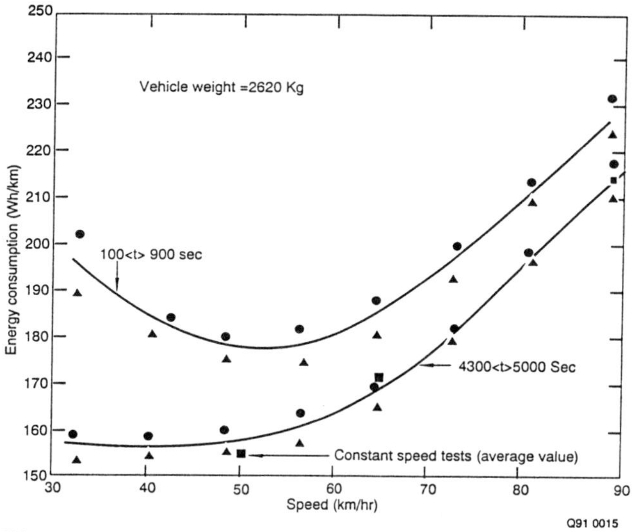

Figure 11. Energy consumption as a function of speed for the DSEP vehicle

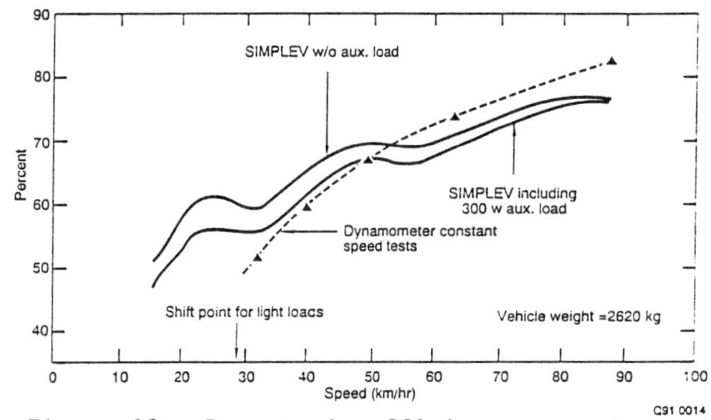

Figure 12. Powertrain efficiency as a function of speed for the DSEP vehicle

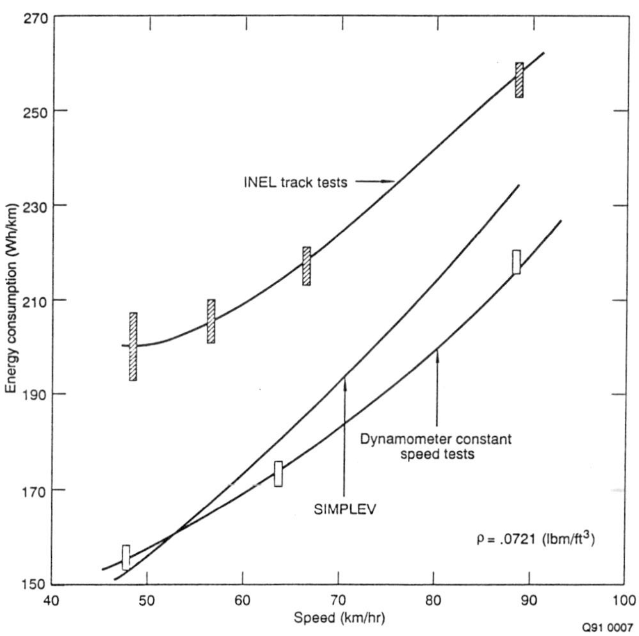

Figure 13. Comparisons of energy consumption values from Dynamometer and track tests with simulation results

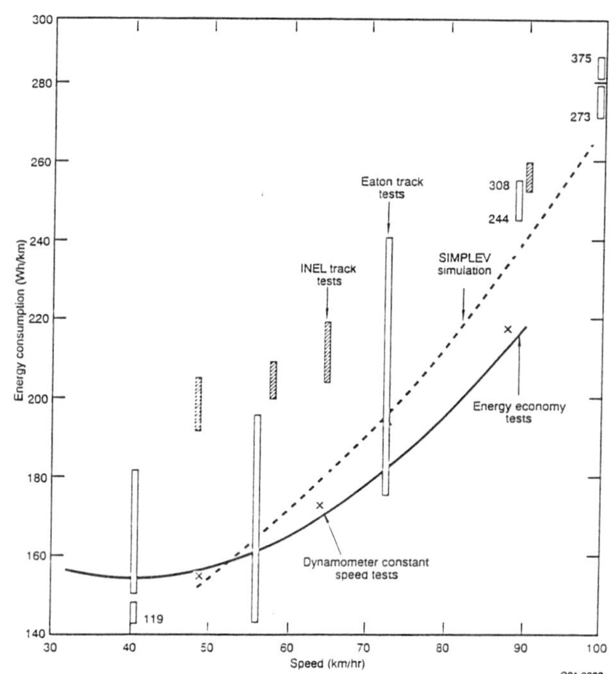

Figure 14. Comparison of energy consumption values from various sources

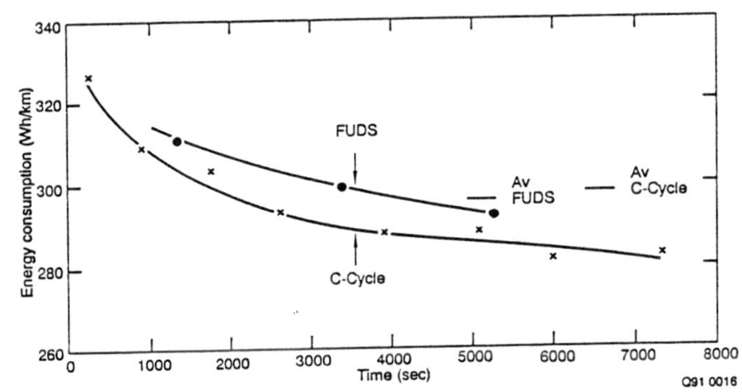

Figure 15. Variation of energy consumption during drive cycle tests on the Dynomometer

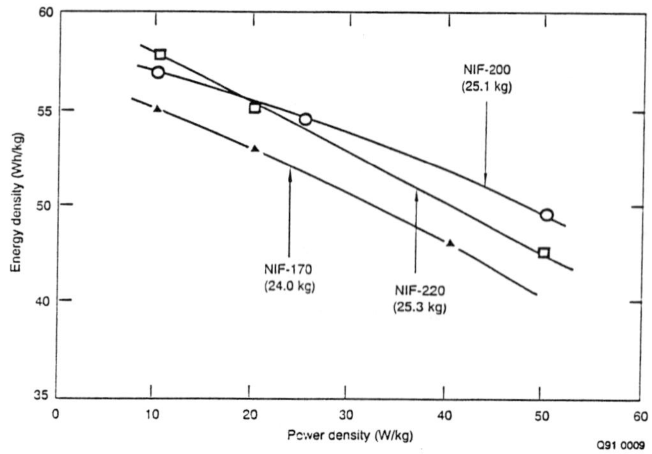

Figure 16. The Ragone Curves for several nickel-iron batteries

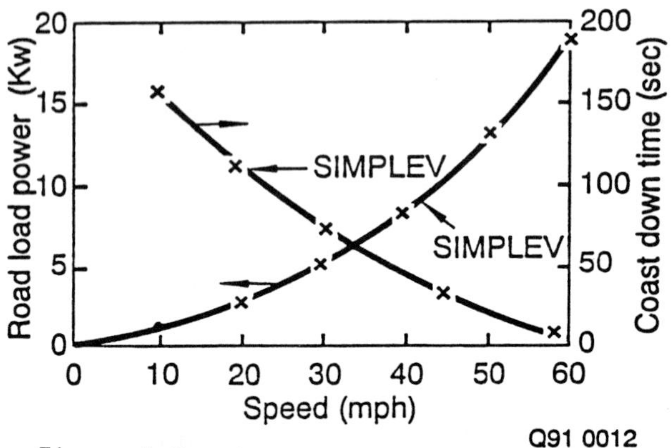

Figure A-2. Comparison of the measured and calculated road load and coastdown characteristics

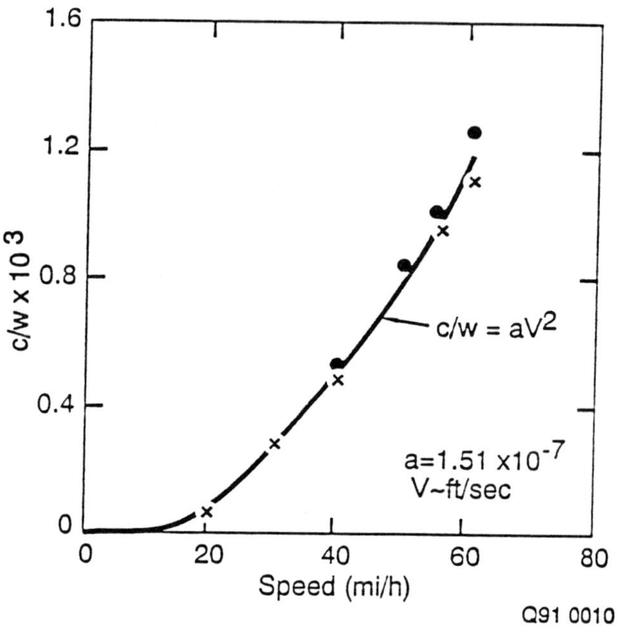

Figure A-1. Variation of the tire rolling resistance coefficient with

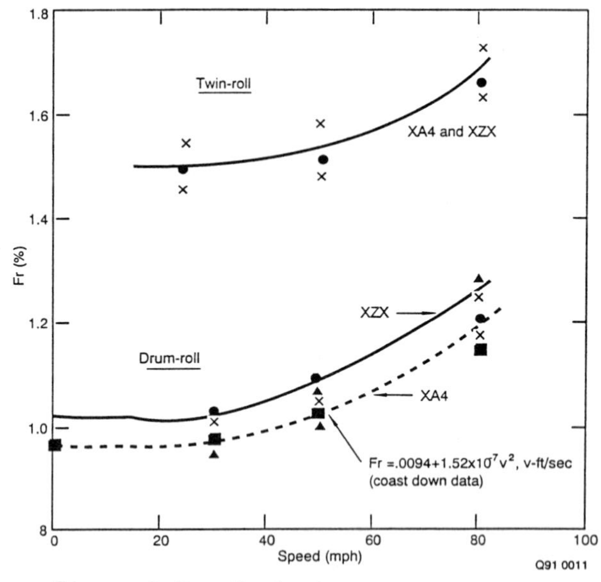

Figure A-3. Variation of tire rolling resistance with speed for several radial tires

910244

Electric Vehicle Development in Fiat

G. Brusaglino
Fiat Central Research Center

ABSTRACT

Within the strategical scope of energy conservation and improvement of human quality of life FIAT is since long time performing activities aimed at the technologies for electrically propelled vehicles for diversified applications.

The electricity as energy vector allows an application flexibility to different types of vehicles, full electrics hybrid and multimode vehicles which have been developed at prototypes level or small scale production.

The tactical process which has been followed is based on the three phases of :

. technology research and development
. demonstration programs and field trials
. production.

The experience achieved within several extensively tested prototypes and pilot service allowed to put in small series production light and medium delivery vans as well as minibuses, which are now in service in Rome and other italian cities.

In 1990, the Panda Elettra has been put in commercialization as a first type of electric vehicle offered for private transport; this vehicle is based on technological solutions adequately and economically viable for the present state of the art.

The evolution process includes now further technological development in the field on board energy sources and drive system.

As a first example of this new technologies, a solution with four "in wheel motor" has been developed and applied to a multimode hybrid bus.

The potential impact, features and issues are discussed for the new technologies which are presently under development.

1. GENERAL FRAMEWORK

The use of electricity as energy vector allows the realization of a variety of vehicular solutions presenting the known advantages in terms of conservation of environmental quality and energetical resources as well as offering possible beneficial features for the structure and management of the vehicle itself.

Fig. 1 shows a general framework diagram of an electrically propelled vehicles in which the two main blocks are put in evidence consisting of the energy supply system and the utilization, which is constituted by the propulsion system and the auxiliaries.

As far as the energy suply system is concerned, two functions can be identified, which can be fullfilled in principle by different subsystems or depending on the application, by the same apparatus :

. an on board energy source facing the overall energy requirement of the vehicle related to the operational mission, and which tipically interfaces the external energy supply source

. a power source, facing the energy peak requirements, which particularily in urban operation characterize the power demand, either in traction or in braking mode.

The power source is tipically to be considered as a functional complement of the energy sources, as required by

GENERAL FRAMEWORK OF ELECTRICALLY PROPELLED VEHICLE

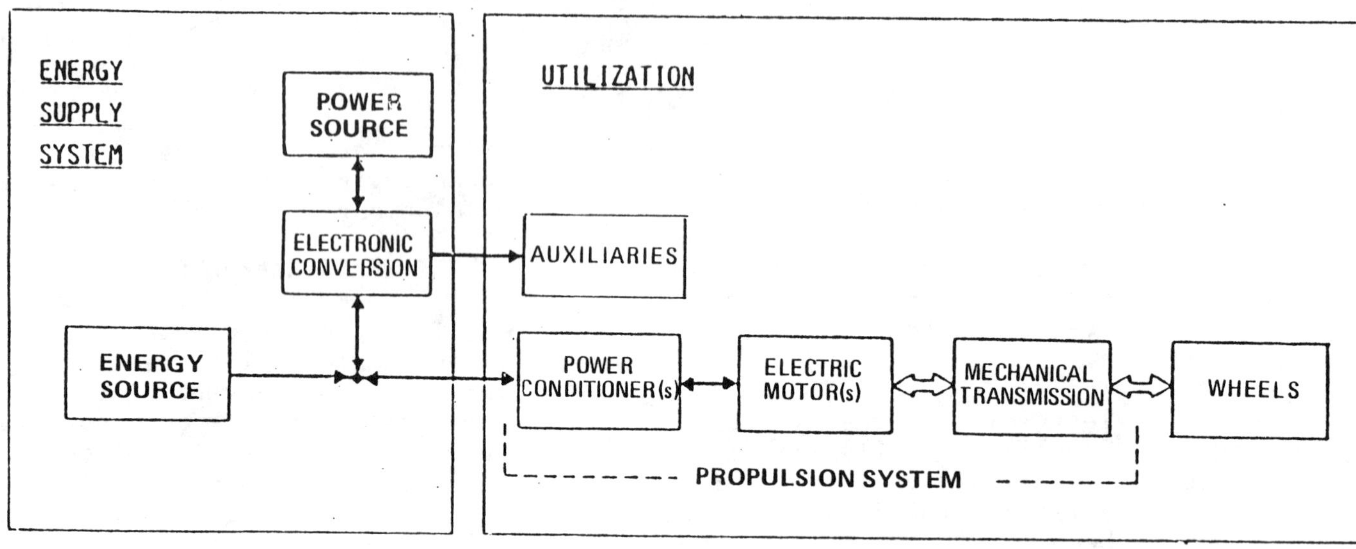

EXTERNAL ENERGY SUPPLY PROCEDURE	ENERGY SOURCE	POWER SOURCE INTEGRATION • ELECTRICAL (BATTERY, SUPERCAPACITOR) • MECHANICAL (FLYWHEEL – MOT. GEN.)
■ FIXED POINT RECHARGING FROM THE ELECTRICAL MAINS	■ STORAGE BATTERY	■ OPPORTUNE FOR PEAK SHAVING AND EFFECTIVE ENERGY RECOVERY
■ CONTINUOUS ENERGY SUPPLY OF ELECTRIC ENERGY	■ EXTERNAL NETWORK	
■ FIXED POINT FUEL SUPPLY	■ HEAT ENGINE – GENERATOR ■ FUEL CELL	■ NECESSARY FOR LOAD LEVELLING AND ENERGY RECOVERY ■ OPPORTUNE FOR OPERATING IN PURE ELECTRIC WITHOUT INFRASTRUCTURE

FIG. 1

the type of the energy source itself and the pattern of utilization.

Fig. 1 also shows a list of possible types of energy sources, which characterize various types of electrically propelled vehicles and require different procedure of external energy supply.

The first two procedures consist of direct electric energy supply from the external mains, either by storing the energy on board while the vehicles is standing idle or by continuous energy supply while the vehicle is running.

The third procedure listed in fig. 1 is based on the concept of converting on board the electrochemical energy of a fuel into electricity by a heat engine-generating unit or a fuel cell. These case allows a quick refueling by an external infrastructure and can also be considered as possible range extender to the first case of pure electric supply mode.
In these cases the integration of the energy source with the power source (which gives place to hybrid system) is to be considered, necessary for achieving a load levelling effect; as far as the heat engine generating unit is concerned, this is beneficial firstly for emission reduction and secondly for fuel economy improvement; in the case of fuel cell it is generally required by its inherent working conditions, including the starting period.

In the cases of the pure electric energy supply the integration of the main energy source with an auxiliary power source can be considered opportune for peak shaving and more effective energy recovery; in the case of battery operated vehicle this can bring to a better overall energy utilization and to wider range, or, over the same mission profile to lower depth of discharge and longer life, and in any case to lower consumption; in the supply by an electrical line this can allow the autonomous on board energy recovery (without requiring a simultaneous presence in line of another vehicle in traction mode), and the down sizing of the electrical line itself to just the average power.

Provided that an adequate energetical capacity is installed, the auxiliary power source can also allow a further mode of operation in pure electric out of the energy supply infrastructure (pure electric dual mode system).
All these features can be particularly suitable, or almost necessary, in the case of energy supply via contactless electromagnetic coupling, for the optimum design of the infrastructure system, under the economical energetical and stand point as well as under the energetical the aspect of minimizing the parasitic field effects.

2. BATTERY VEHICLES DEVELOPMENT : FROM RESEARCH TO PRODUCTION

Since years '60s FIAT has been conducting research and development on electric propulsion technology, already at that time under the motivation of environmental protection (fig. 2).

In the years '70s another motivation has been added related to conservation of energy resources.

In a first period the activity was dedicated to development of propulsion systems based on techniques meeting the requirements of easy driveability and energy saving (e.g. control strategies, suitable for urban operation, including regenerative braking).

The storage system was at the same time developed in terms of optimum use of the batteries, including thermal management, peripheral devices and safety feature.

The activity was then devoted to the integration of such systems into the vehicular structure.

In that framework a concept car, the X1/23, was realized in the years '72-'74, according to a criterion of functional integration of the developed thechnologies into a vehicle structure ad hoc conceived to meet the dimensional and manoeuvering requirements for the urban use.

The attention was then dedicated to the commercial vehicles and vehicles for public service; actually in these

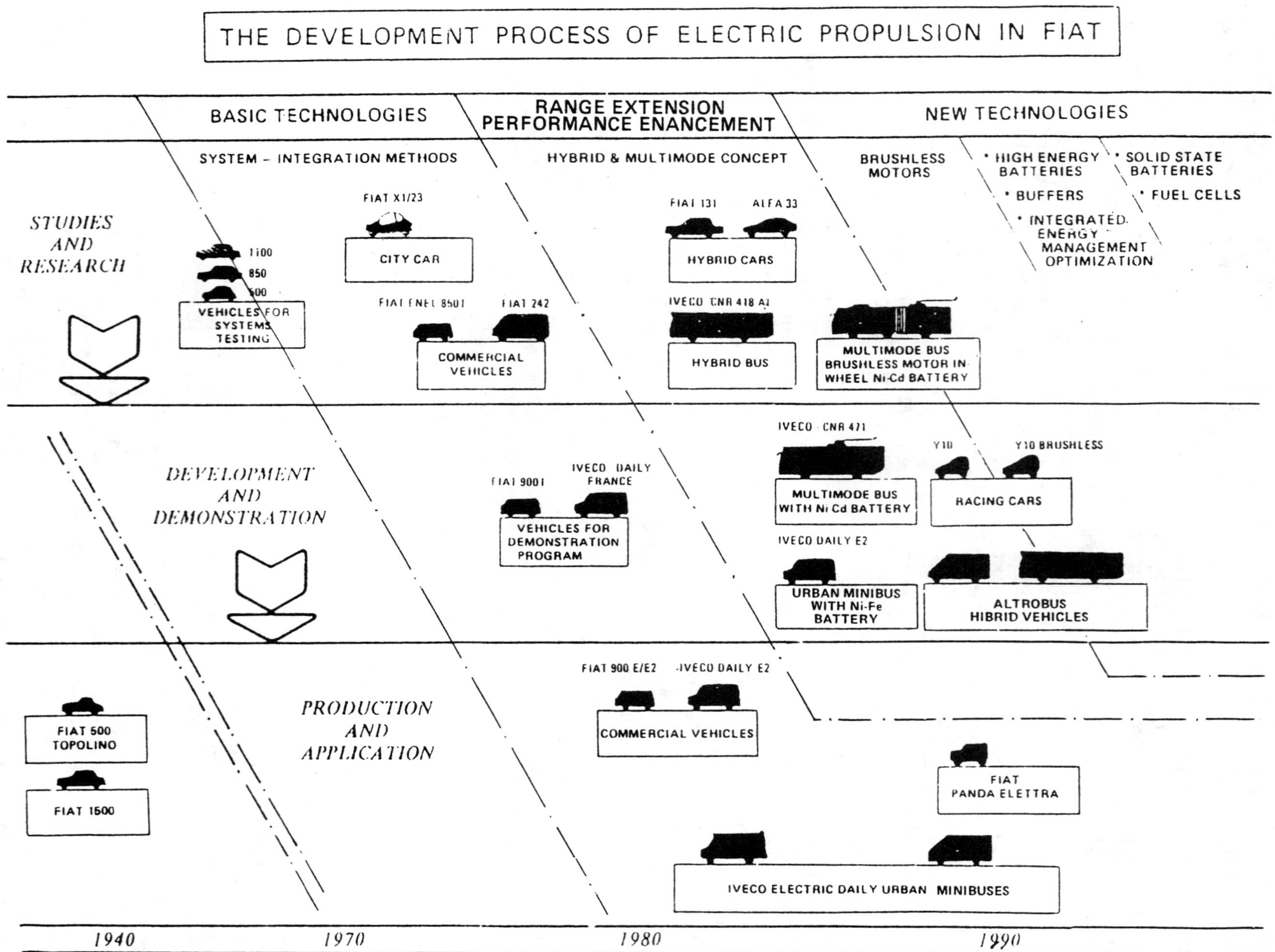
THE DEVELOPMENT PROCESS OF ELECTRIC PROPULSION IN FIAT
BASIC TECHNOLOGIES
RANGE EXTENSION PERFORMANCE ENANCEMENT
NEW TECHNOLOGIES
SYSTEM - INTEGRATION METHODS
HYBRID & MULTIMODE CONCEPT
BRUSHLESS MOTORS
• HIGH ENERGY BATTERIES
• BUFFERS
• INTEGRATED ENERGY MANAGEMENT OPTIMIZATION
• SOLID STATE BATTERIES
• FUEL CELLS
STUDIES AND RESEARCH
1100
850
500
VEHICLES FOR SYSTEMS TESTING
FIAT X1/23
CITY CAR
FIAT ENEL 850T
FIAT 242
COMMERCIAL VEHICLES
FIAT 131
ALFA 33
HYBRID CARS
IVECO CNR 418 AI
HYBRID BUS
MULTIMODE BUS BRUSHLESS MOTOR IN WHEEL Ni-Cd BATTERY
DEVELOPMENT AND DEMONSTRATION
FIAT 900 I
IVECO DAILY FRANCE
VEHICLES FOR DEMONSTRATION PROGRAM
IVECO CNR 471
MULTIMODE BUS WITH Ni Cd BATTERY
Y10
Y10 BRUSHLESS
RACING CARS
IVECO DAILY E2
URBAN MINIBUS WITH Ni-Fe BATTERY
ALTROBUS HIBRID VEHICLES
FIAT 500 TOPOLINO
FIAT 1500
PRODUCTION AND APPLICATION
FIAT 900 E/E2
IVECO DAILY E2
COMMERCIAL VEHICLES
FIAT PANDA ELETTRA
IVECO ELECTRIC DAILY URBAN MINIBUSES
1940
1970
1980
1990

FIG. 2

categories of vehicles and in their use in urban areas the earliest opportunities of electric propulsion application were identified.

Various prototypes were developed and small series produced for demonstration programs and pilot services.

In this field mention has to be made for the definition of the first generation of the van Fiat 850T which has been done in collaboration with ENEL, the Italian National Electric Utility.

From this model a series of 20 electric van Fiat 900T was derived for first trials in the fleets of ENEL and SIP (the Italian Telephone Company).

A further model in the field of light-medium vans was the IVECO Daily electric; this vehicle was developed for the international bid issued by the French "Group Interministeriel Vehicule Electrique", aiming at the definition of electric vehicles for French Utilities and Public Organizations, according to assigned specifications.

The IVECO Daily electric van, after a qualification testing programe performed by Electricite' de France, was the vehicle selected in the category of 1.000 Kg payload.

On the basis of the experimental results of these opeations, which produced a first technological response and on the basis of the application potential, which in the mean time were preliminarily identified, a small series production was initiated in 1982 of a second generation of light-medium vans: the FIAT 900E/E2 (fig. 3) and the IVECO FIAT Daily E2 (fig. 4); they feature driving characteristics and performance meeting the requirements of urban operation, with a range of 60-80 Km in city driving pattern and up to 120 Km at constant speed of 50 Km/h.

The development of the second generation of vans was based on the concept of achieving also acceptable overall cost even in production in very small series, and to make the operation of the vehicle easy and effective from the user standpoint.

The design of the vehicles provided limited modifications of the body of the conventional vehicle and these were introduced already in the production line of the conventional vehicle itself.

The electrical components were conceived in modular way for both types of vehicles, for achieving a better effect of scale.
The concept was to work at the same level of current, and increrasing the power, according to the different size of vehicles, acting on the voltage; this alloxed to use the same battery modules, the same cross section of the electric motors and the same components for the chopper.

Figg. 5 and 6 show the d.c. separately excited motors of these vans as well as the measured characteristics of the propulsion system; notice the symmetric behaviour between traction and braking and the slightly increase of efficiency going to higher voltage.

Fig. 7 shows the related double chopper.

Some 20 vehicles of these series were selected by the Commission of the European Communities for demonstration programs to be held in the first half of years '80s in three european cities: Dublin, Odense and Torino /1/.

In the same period some 12 units of these series were protagonists within a demonstration program sponsored by theItalian National Research Council in torino, where electric vehicles of different type were put in service in the fleed of Public Utilities and Municipal Authorities for a validation of their operational capabilities and an assessment of their energetical performance /2/.

The development of the second generation of vans was based on the concept of achieving also acceptable overall cost even in production in very small series, and to make the operation

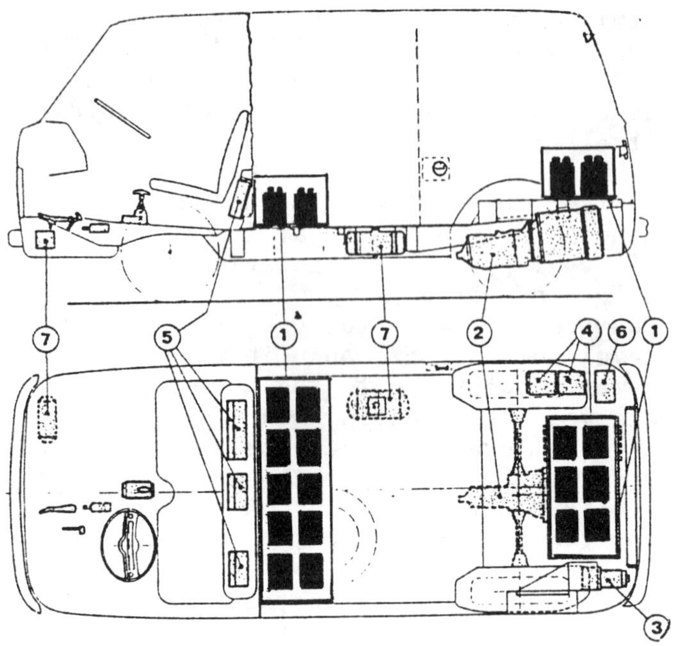

FIG. 3 - BASIC LAYOUT AND CHARACTERISTICS OF THE FIAT 900E/E2 ELECTRIC VAN

1. traction battery
2. drivertrain
3. electric van for motor cooling
4. chopper
5. battery charger
6. auxiliary battery
7. heater

Range in urban cycle	60÷80 km
Payload	430 kg
Maximum speed	60 km/h
Range at constant speed (50 km/h)	90 km
Acceleration: 0--> 50km/h	12 s
Curb weight	1470 kg
Motor: type	DC separately excited
nominal power	96 V
Battery : type	Pb
nominal voltage	96 V
weight	480 kg

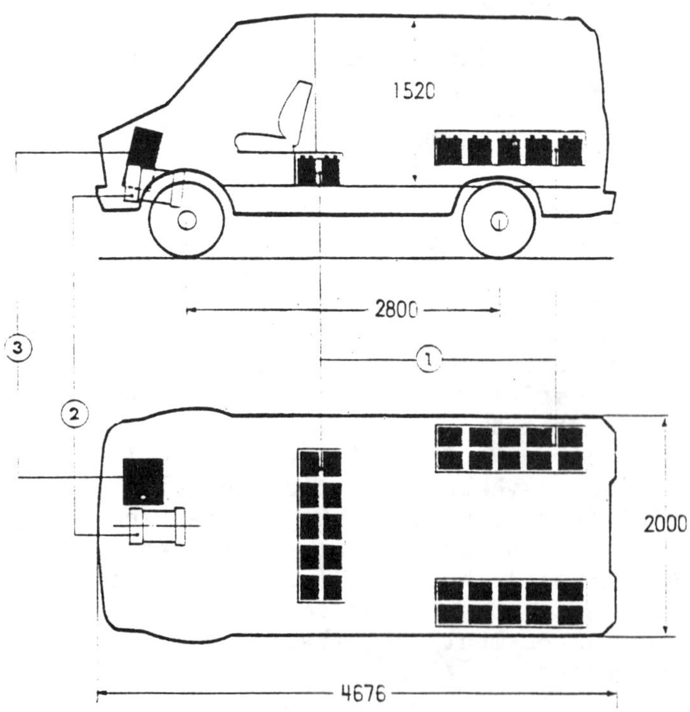

FIG. 4 - IVECO FIAT DAILY E2

1. traction battery
2. traction motor
3. chopper

Range in urban cycle	80 km
Payload	1000 kg
Maximul speed	65 km/h
Range at constant speed (50 km/h)	120 km
Acceleration: 0-->50 km/h	14 s
Curb weight	2800 kg
Motor : type	DC separtely excited
nominal power	20 kW
Battery : type	Pb
nominal voltage	180 V
weight	980 kg

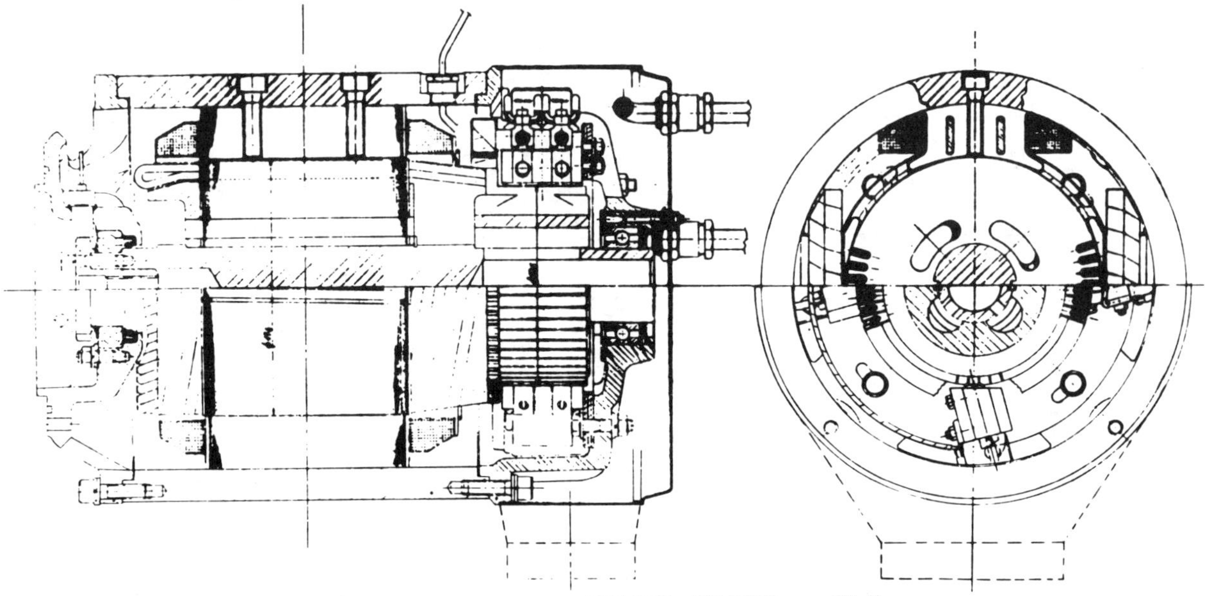

NOMINAL VOLTAGE : 96 V
NOMINAL CURRENT : 150 A **PEAK CURRENT : 400 A**
NOMINAL POWER : 12 kW **PEAK POWER : 26 kW**

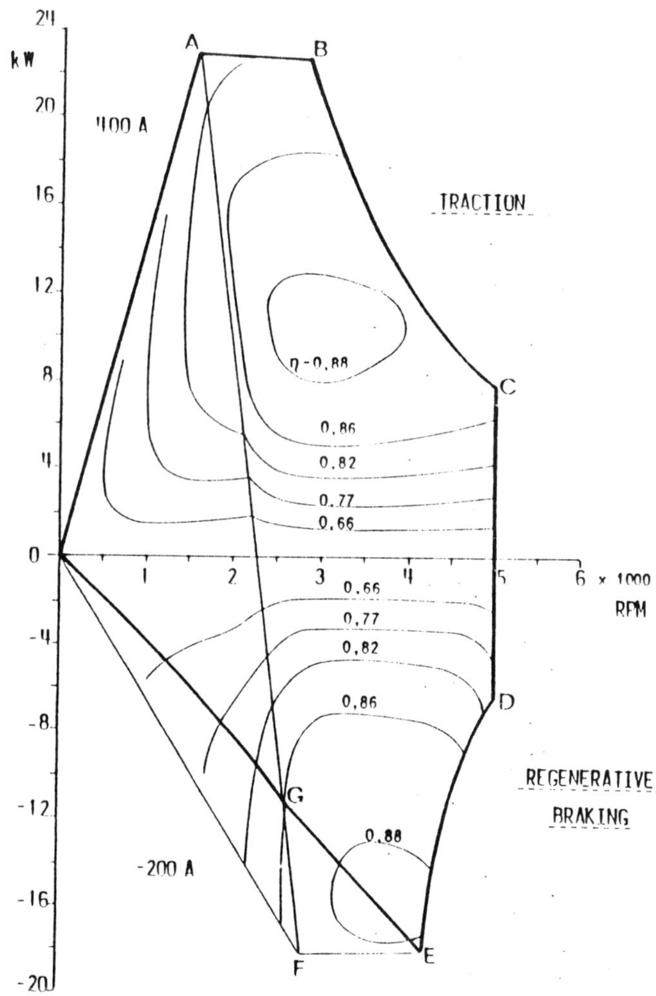

FIG. 5 - TRACTION MOTOR CHARACTERISTICS OF FIAT 900E/E2 ELECTRIC VAN

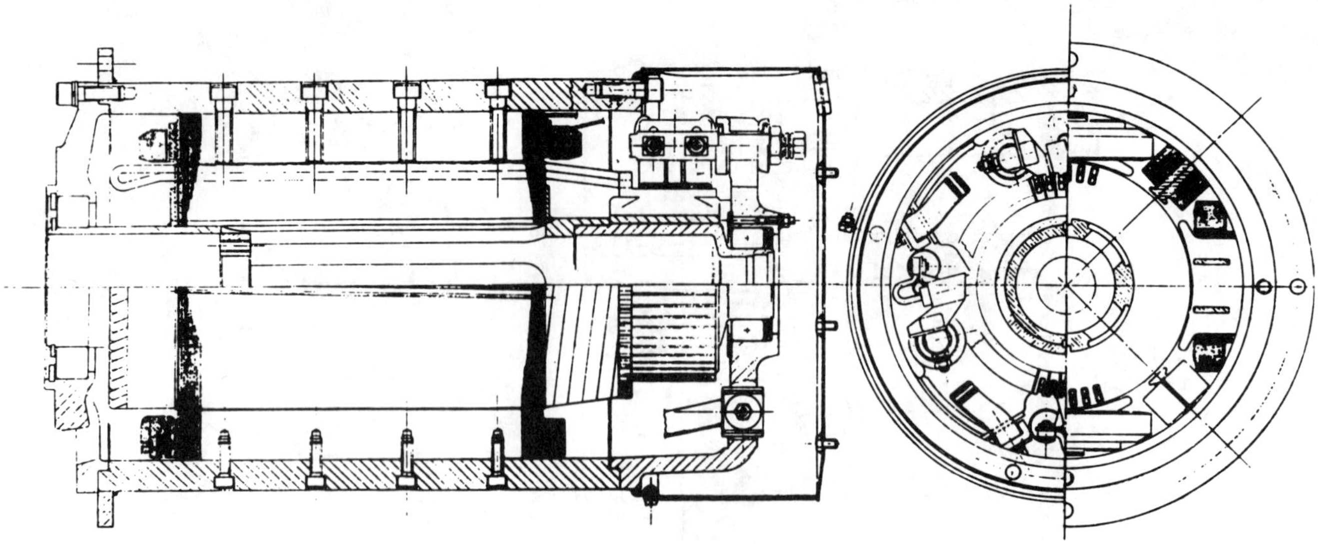

NOMINAL VOLTAGE : 180 V
NOMINAL CURRENT : 150 A   **PEAK CURRENT : 400 A**
NOMINAL POWER : 23 kW   **PEAK POWER : 47 kW**

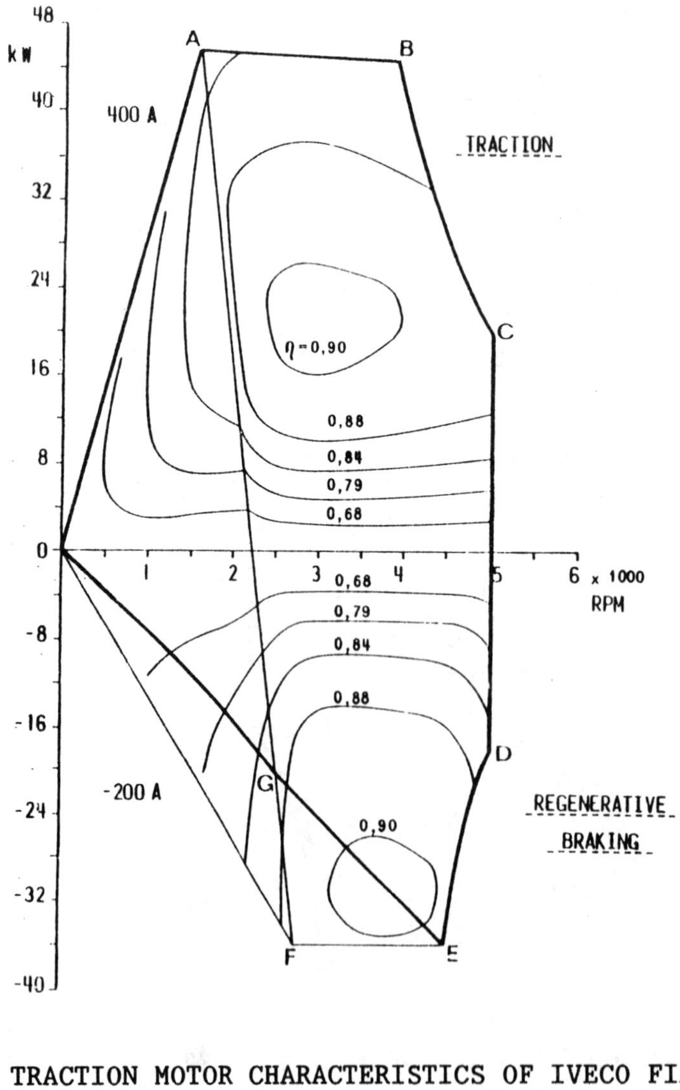

FIG. 6 - TRACTION MOTOR CHARACTERISTICS OF IVECO FIAT DAILY E2

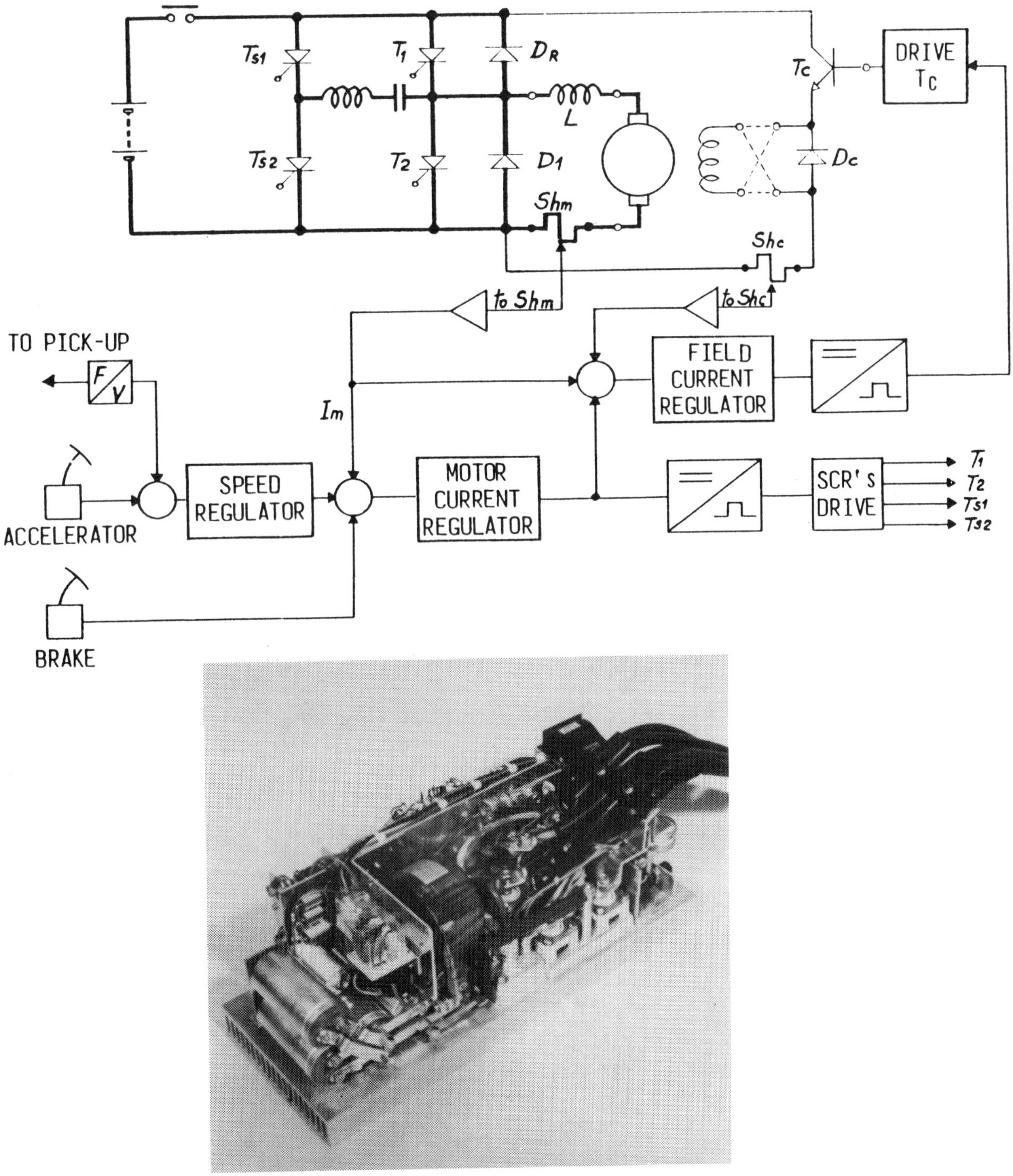

FIG. 7 - DOUBLE CHOPPER FOR SEPARATELY EXCITED MOTORS FOR FIAT 900E/E2, IVECO FIAT DAILY E2 VAN AND MINIBUS

Range in urban cycle	60÷70 km
Payload	16÷20 passengers
Maximum speed	60 km/h
Range at constant speed (50 km/h)	80÷90 km
Acceleration : 0-->30 km/h	7 s
Curb weight	3900 kg
Motor : type	DC separately excited
nominal power	22 kW
Battery : types	Pb-NiFe
nominal voltage	192V 192 V
weight	960(Pb) 652 Ni-Fe kg

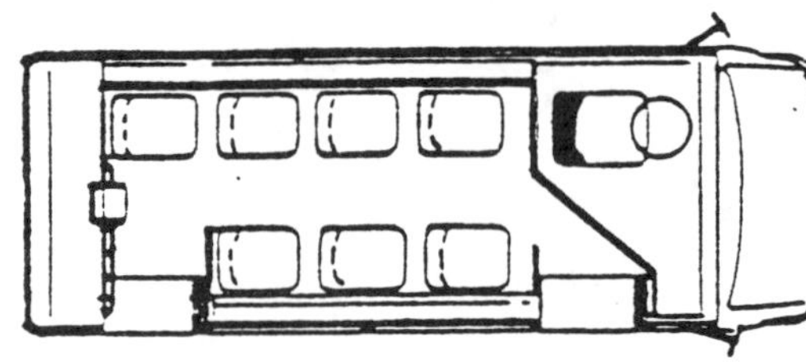

FIG. 8 - IVECO DAILY MINIBUS CHARACTERISTICS

FIG. 9 - IVECO DAILY MINIBUS WITH NEWLY DESIGNED BODIES

In the field of passenger cars, in the spring of 1990 the Fiat Panda Elettra has been put in commercialization.

Fig. 10 and 11 show views and main characteristics of the Panda Elettra.

The criteria which has been followed in the vehicle design and the definition of the board system is consistent with the general concept of offering to the customer consolidated technical solutions with the same level of safety and comfort of conventional vehicles, as well as comparable driving performance in urban city, featuring easy and totally traffic compatible mobility, on level or hilly roads.

3. HYBRID AND MULTIMODE VEHICLES

Another approach to overcome the range limits allowed by the presently available storage batteries other thanthe already mentioned interim recharging, is the hybrid solution in which the propulsion system is supplied with energy by an integrated system consisting of two sources, a storage unit (typically an electrochemical storage battery) and an other energy source featuring quick refueling possibility (typically a heat engine).

An experimental development of a hybrid bus has been done by Centro Ricerche Fiat in the framework of the Finalized Project Energetics of the Italian National Research Council (CNR) in the year '78-'81, according to the concept of series hybrid system: diesel generator - lead acid battery - electrical transmission - propulsion systemwith d.c. separately excited motor /3/.

The battery can assure the pollution free operation in "pure electric" mode in the city centers, or, more in general, in the environmentally protected areas; but also in "hybrid mode" operation, with heat engine on, the system allows the improvement of its working conditions from the pollution standpoint and also under the consumption aspect.

Actually the battery releases the power supply pattern of the heat engine from that required by the propulsion system; covers the peak power need during acceleration and recovers energy during braking.

This allows the installation of a heat engine of reduced power, its optimization in terms of emissions and consumption and easier reduction of noise, due to the absence of mechanical connections with the vehicle structure. Fig. 12 shows the basic layout of the vehicle.

The hybrid bus CNR-IVECO 418 AL has proven within practical tests in Torino a reduction of fuel consumption of more than 20% in comparison with a corresponding conventional bus operating on the same loading and mission conditions. The concept has been further developed in a recent realization, the "Altrobus" in the framework of a cooperation IVECO-FIAT, Ansaldo, Genova, Ricerche, with a more specific finalization for emission reduction.

Fig. 13 and 14 show a first prototype derived from IVECO FIAT Daily, and fig. 15 a further prototype, presently under testing equipped with asyncronous motor and Ni-Cd batteries.

A further solution consists on the dual-mode vehicle, in which two energy supply modes are provided: one typically "pure electric" and the other one, to be use as an alternative, which features quick refueling possibility (typically a heat engine).

As a conceptual difference between dual-mode and hybrid , it is intended that in the dual-mode system the two energy sources work alternatively, at different time, while in the hybridsystem the two energy sources work in an integrated way.

A possibility, conceiveable for passenger cars or commercial vehicles is to introduce, in addition to the conventional heat engine, an electric propulsion system with battery (which, at its turn can be recharged by the mains or by the heat engine while

FIG. 10 - FIAT PANDA ELETTRA

PANDA ELETTRA

- STORAGE BATTERY
  12 Pb 6V modules in series (72V total)
  total weight : 350 kg
- ELECTRIC MOTOR
  d.c. series excited
  nominal power : 9.2 kW
  manual shift 4 speed + reverse gear
- RANGE
  on urban ECE cycle: 70 km
  at 50 km/h constant: 100 km

  with option Ni-Cd battery:
  on urban ECE cycle: 100 km
  at 50 km/h constant: 180 km

- MAXIMUM SPEED
  70 km/h
- ACCELERATION
  0 - 40 km/h in 10 s
- CLIMBABILITY
  25%
- PAYLOAD
  2 persons + 100 kg

- BATTERY CHARGER : ON BOARD, AUTOMATIC, 220 V, 16 A
  RECHARGING TIME 8 h
- PASSENGER COMPARTMENT HEATER : GASOLINE HEATER
- AUXILIARY SYSTEM : 12 V BATTERY FED BY DC/DC CONVERTER
- CURB WEIGHT : 1050 KG
- REGENERATIVE BRAKING : AUTOMATIC

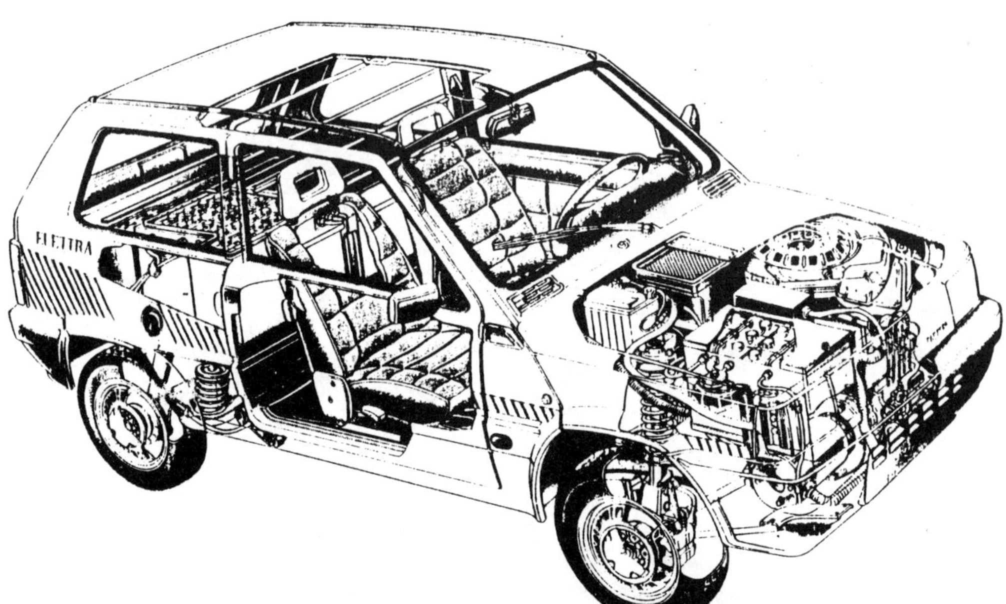

FIG. 11 - CHARACTERISTIC AND PERFORMANCE OF FIAT PANDA ELETTRA

HYBRID BUS WITH STORAGE BATTERIES
COMPONENTS LAYOUT

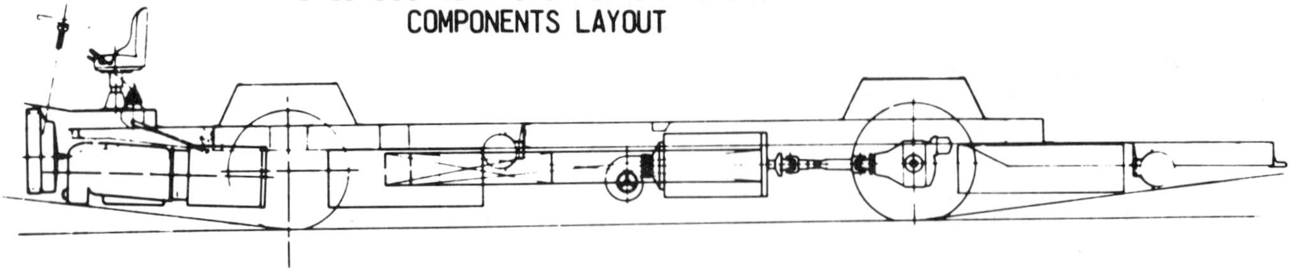

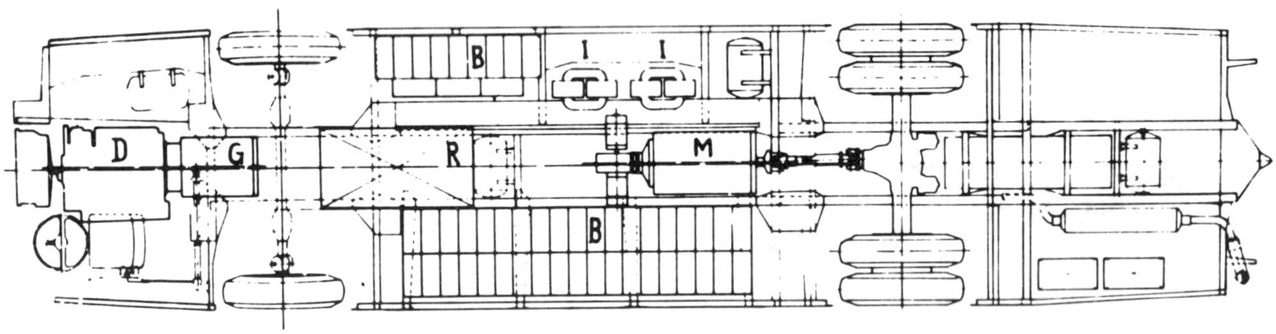

D - DIESEL ENGINE 6 CILINDER FIAT 80 60.12.000

G - D.C. GENERATOR

M - D.C. TRACTION MOTOR

B - LEAD STORAGE BATTERIES 600V 135 Ah (5 h)

R - ARMATURE AND FIELD POWER CONDITIONER

I - SMOOTHING INDUCTANCES

FIG. 12 - BASIC LAYOUT OF THE IVECO-CNR 418AL HYBRID BUS

FIG. 13

FIG. 14

PROTOTYPES OF HYBRID BUSES "ALTROBUS"

operating beyond the environmentally restricted areas).

A solution typically suitable for buses, is to consider as alternative energy sources an electric trolleybus line and a generating unit using fluid fuel, typically a heat engine or a fuel cell.

As an example of a practical achievement, in the framework of the Finalized Project Transport of Italian National Research Council (CNR), Centro Ricerche Fiat, on behalf of A.T.M. of Milano (the Municipal Transport Authority) and with the support of IVECO has developed a dual-mode bus derived from the model IVECO FIAT 471 /4/, /5/, /6/, /7/.

In this realization, which involved also Ercole Marelli Trazione and SEAC-Viberti, teh followed approach consists on the integration od the system for operation under trolley bus line with a diesel-generator unit and

In this realization, which involved also Ercole Marelli Trazione and SEAC-Viberti, the followed approach consists on the integration of the system for operation under trolley bus line with a diesel-generator unit and with a high power Nickel-Cadmium battery, according to the series hybrid concept.

Fig. 15 shows the basic scheme of the system and the main characteristics of the vehicle.

The two modes of energy supply are thus a trolleybus line and a hybrid system diesel-generator with battery.
The high power Ni-Cd battery is involved in the operation also in the electric line supply mode, thus giving place also in this mode to a hybrid system; the advantages, aleady mentioned in the chapter 1 are the peak shaving from the line and the autonomous on board energy recovery.

Finally the battery allows a third mode of energy supply in pure electric without infrastructure; thus the vehicle can be definede as "multimode" and allows a great flexibility of operational management as well as in terms electric line installation and use.

On the other hand, these various mode of enegy supply and operation are obtained with a high degree of integration and utilization of the on board elements, as put in evidence in the block diagram of fig. 15.

4. TECHNOLOGICAL DEVELOPMENT

As a parallel activity of application of the achieved technologies to vehicles and their operational use, research and development lines have been initiated aimed at the improvement of the performance.
Some outline consideration are below reported concerning the area of action which are also in line with the technological impact study performed within the European Action COST 302, developed from 1981 and 1987 /7/.

- Energy supply systems -

With reference to the scheme of Fig. 11, the following energy sources and power sources are considered :

Energy sources	Power sources
- Storage batteries	- Electrical storage
. Lead-acid . Nichel-alcaline . Zinc-Bromine . Sodium-Sulphur . Lithium solid state	. Lead-acid . Nickel-Cad. . Supercapac.
- On board fuel conversion	- Mechanical storage
. Fuel cells	. Flywheel motor-gener.
. Heat engine-generator unit	

Prospect reported at Fig. 16 summarizes the situation of the on board energy supply system, in terms

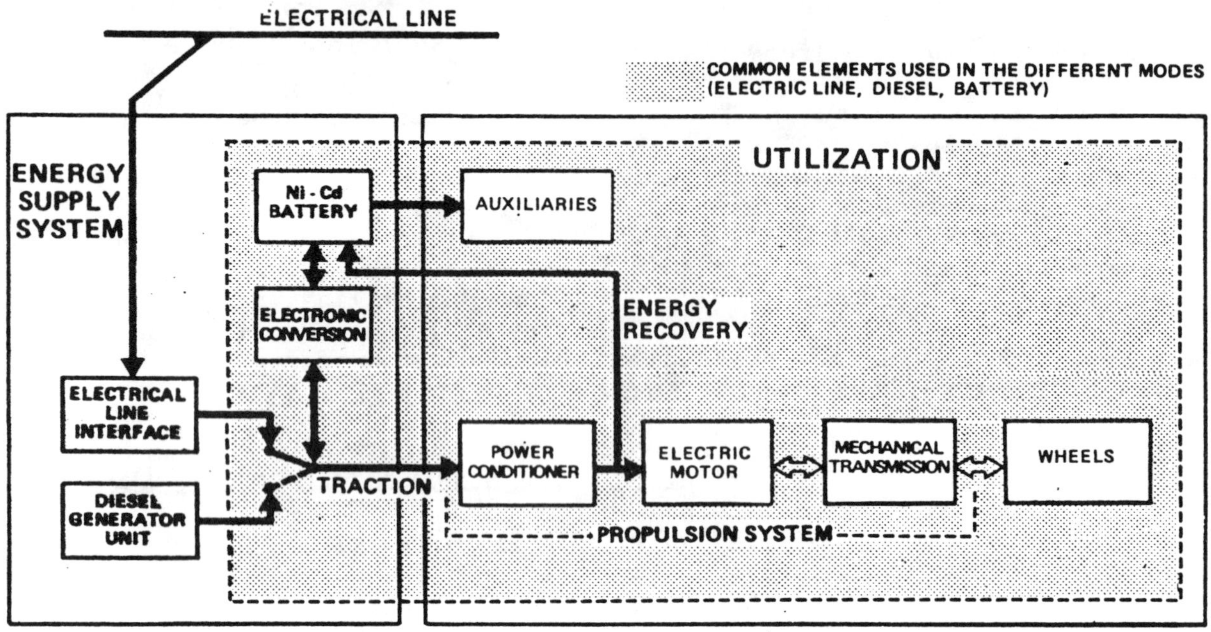

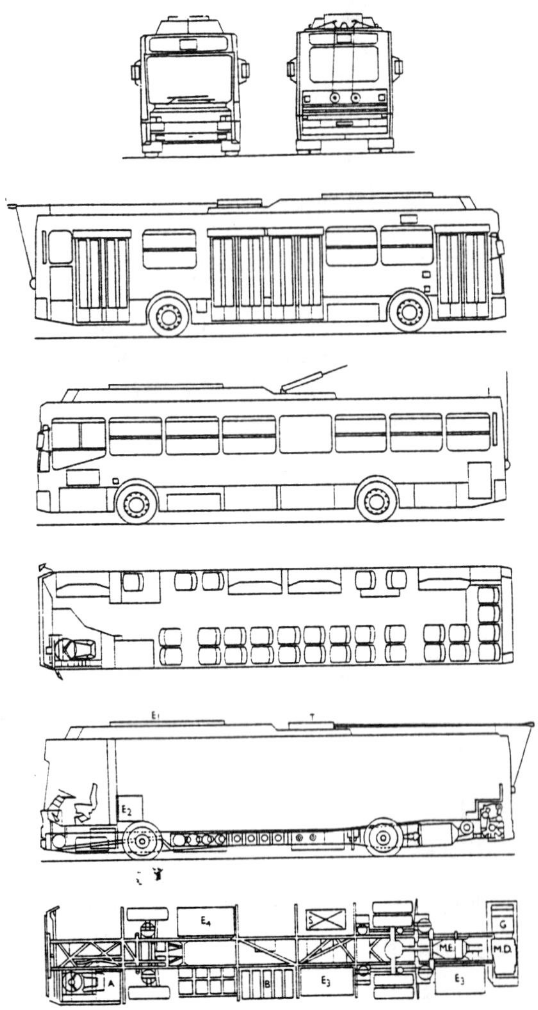

Payload	140 persons
Maximum speed	65 km/h
Acceleration	1 m/sec
Cub weight	19000 kg
Motor : type	DC separatey excited motor
Nominal power	90 kW
Battery : type	Ni/Cd
nominal voltage	240 V
weight	400 kg

E = ELECTRIC EQUIPMENT/ELECTRONIC CONTROL
MD = DIESEL ENGINE
G = GENERATOR
B = Ni-Cd BATTERY
ME = ELECTRIC MOTOR

FIG. 15 - DUAL MODE VEHICLE IVECO-FIAT CNR

	STORAGE BATTERIES					FUEL CELLS	
	Pb	Pb recombination	Ni-Cd	Na-S	Li-SS Solid state	Fluid fuel	Solid fuel (Al-air)
FEATURES	• WIDELY USED (E.G. FORK, LIFTS, INDUSTRIAL TRUCKS)	• MAINTENANCE FREE	• SPECIFIC ENERGY TWICE THE Pb • HIGH POWER	• MAINTENANCE FREE	• SPECIFIC ENERGY 5 TIMES THE Pb • MAINTENANCE FREE • SHAPE FLEXIBILITY	• QUICK REFUELING	
AVAILABILITY		1990		1994	1998		
PROBLEMS/ DEVELOPMENT NEEDS	• NEED OF WATER REFILLING • LOW SPECIFIC ENERGY	• SPECIAL RECHARGING PROCEDURE	• PROBLEMS OF PRODUCTION AND RECYCLING DUE TO CADMIUM • HIGH COST	• THERMAL MANAGEMENT • INDUSTRIAL DEVELOPMENT	• NEED OF RESEARCH AND DEVELOPMENT	• NEED OF RESEARCH AND DEVELOPMENT • NEED OF INTEGRATION WITH BUFFER • SYSTEM MANAGEMENT	

FIG. 16 - ON BOARD ENERGY SUPPLY SYSTEMS

of features and problems-necessity of development according to FIAT view point.

Fig. 17 shows a more detailed assessment electrochemical energy supply systems, storage batteries and fuel cells, cosidered in scale production in comparison with lead-acid batteries.

- <u>Propulsion systems</u> -

The present technology of propulsion systems for battery operated EVs is mainly based on the d.c. electric motor (separately excited for the majority of applications), whose mechanical performance is electronically controlled by the so called "chopper". This type of propulsion system, provided appropriate design and manufacturing, allows satisfactory performance, smooth acceleration andregenerative braking. However, new technological approaches are in the research and development stage, having the aim of achieving improvements in one or more of the following features :

	Pb	Pb RECOMBINATION	Ni-Cd	Na-S	Li-V_6O_{13}	FUEL CELLS SOLID STATE
SPECIFIC ENERGY (Wh/kg)	25 ÷ 35	SAME ORDER	TWICE	FOUR TIMES	FIVE TIMES	ORDER OF 10 TIMES (DEPENDING ON FUEL TANK)
SPECIFIC POWER (W/kg)	80 ÷ 120	SAME AS Pb	FOUR TIMES	SAME AS Pb	HIGHER	REQUIRES BUFFER
ENDURANCE (operating cycles)	500 ÷ 800	LOWER	THREE TIMES	SAME ORDER	SAME ORDER	TIME DEPENDANT
WORKING TEMPERAT. (C)	-20 ÷ 60	SAME AS Pb	SAME AS Pb	320	120 AMBIENT	AMBIENT HIGHER FOR THE REFORMER
REFUELING (h)	6 ÷ 8	SAME ORDER	HALF	ONE QUARTER	HALF	
COST (ECU/Wh)	0.1	SAME ORDER	THREE TIMES	SAME ORDER	SLIGHTLY HIGHER	
MAINTENANCE	WATERING	NONE	WATERING	NONE	NONE	
STATUS	AVAILABLE	AVAILABLE	AVAILABLE	PROTOTYPES	IN DEVELOP.	IN DEVELOPMENT

FIG. 17 - PREDICTABLE FEATURES OF ELECTROCHEMICAL ENERGY SUPPLY SYSTEMS CONSIDERED IN SCALE PRODUCTION COMPARED WITH LEAD ACID BATTERIES

- weight and dimensions
- flexibility of design and application
- efficiency
- cost.

Systems which appear to have an impact on the above mentioned features are those based on commutatorless motors fed by electronic power conditioners which incorporate the functions of cyclic commutation on the armature windings and control of the mechanical performance.

In this framework, on the basis of the experience of manufacturers and research operators involved in the field of electric propulsion, an analysis has been carried out, in comparison with the system based on the d.c. motor, for the systems based on the following motors :

- asynchronous (induction)
- synchronous electrically excited (wound motor)
- synchronous, excited by permanent magnets (Sm-Co or Nd-Fe-B, or new high performance permanent magnets presently under research-development).

All these motors are fed by inverters, which can be characterized by a substantially similar structure of the power electronics and a control logic depending on the particular type of motor.

In the framework of the European Action COST 302, a comparative analysis has been performed with reference to three values of nominal power of the motor, respectively 10, 15 and 40 kW (suitable for the already mentioned three categories of vehicles). The results can be summarized as follows.

The weight of the commutatorless motor is much lower than the weight of the d.c. motor : the ratio is in the range of 0.4÷0.5 . This a consequence of the absence of the mechanical commutator and, even more important, of the absence of the limits for the rotational speed and/or for the current density, which are tied to the mechanical commutation. As far as the controller itself is concerned, for the present state od the art, the weight of the controller for the commutatorless motor is higher than the controller for the d.c. motor (the ratio can be some 1.5); but this is referred to a weight which is of relatively low importance in comparison with the total weight of the system, so that a considerable advantage still exists for the commutatorless solutions, considering the total system.

The weight ratio between the total system with commutatorless motors and the total system with the d.c. motor ranges from 0.5 and 0.7, depending on the type of system and the category of vehicle.
Configurations with high current density can take advantages also from the possibility of an easier cooling (eventually by liquid) of the armature windings and the parts of the magnetic circuit with variable flux, which are located in the stator. Future technological development mainly in the field of electronics and in that of the permanent magnets, will predictably bring to further reduction of weight, more for the commutatorless motor systems than for the d.c. motor system.

The absence of the mechanical commutator allows a large design flexibility, both in terms of shape and of possible trade-off between the parameters of the power (torque and speed), which can be defined according to the specific application. It is in principle possible to range from motors designed for very high speed to motors designed for very high torque at low speed.

The systems with commutatorless motors present a slightly better efficiency in comparison with d.c. motor; this in particular for the synchronous motors, both with electromanetic and permanent magnetexcitation : for these types of systems, the improvement of efficiency is of the order of 5-6% around values of 85%, and, furthermore, the variations with the output power and speed are lower.

In terms of economics, commutatorless motors systems are in general using, to different extent, more expensive electronics, but a certain degree of compensation is found by the relatively lower cost of the motors themselves, due to their lower weight and simpler construction. According to the comparative evaluation between each commutatoless motor system and the d.c. motor plus chopper, on the basis of the basis of the present technology for all systems, it has been estimated that :

- The system consisting of asynchronous motor plus inverter is in the same range of cost as the d.c. system.

- The system consisting of synchronous motor, electrically excited, plus inverter, is more expensive, the ratio ranging from 1 to 1,2 depending on the class of vehicle.

- The system consisting of synchronous motor, excited by permanent magnets, is at the same level of cost of the d.c. systems or a slightly more expensive (the ratio ranging from 1 to 1,2) considering the Samarium-Cobalt magnets. If Neodymium-Iron-Bore magnets were used, which are now under development, the cost of the motor could be slightly reduced.

In the analysis, the future technology is taken as a reference, according to the predictable developments in a time horizon in which, these new systems could be sufficiently validated, the comparative evaluation goes in the favourable direction for the systems with commutatorless motors.

This is due to the predictable future reduction of cost of the electronics (both power electronics and micro-electronics) as a difference withelectromechanical parts, considering the predominant role of the electronics in the cost figure of the systems with commutatorless motors.

As a matter of fact, the ratio between the cost of electronics and the total cost of the systems, according the performed evaluation, ranges from 0.8, for the system with asynchronous motor, to around 0.67 for the system with synchronous motors, while for the system with d.c. motor it is lower than 0.5.

As a conclusion, in parallel with the application of the d.c. motor system to the electric vehicles of thepresent generation, the technology of the systems with commutatorless machines appears to be worthy of development and application, for its technical benefits. In terms of costs the d.c. motor systems can be considered in most cases the chapest solution at present time, but the competitivity of systems with commutatorless machines is expected to increase with the future development of the electronics, so that this technology could be the best one for future EVs .

Prospect reported at Fig. 18 summarizes the situation of the traction motors and related controllers in terms of features and problems/development needs.

An assessment of the potential impact of high performance permanent magnets on the elctric motor features has been done in the past with the design and testing of a d.c., 8 poles Sn-Co magnets , mechanically commutated disk motor.

Fig. 19 shows the disassembled motor, featuring ironless rotor. It has been tested that by use of this type of excitation a specific torque double than that of a conventional ad hoc designed separately excited d.c. machine can be achieved.
The following step was the done with a brushless electronically commutated and controlled permanent magnet motor integrated in wheel for an electrically propelled bus (fig. 20). The motor was designed and built, together of the related electronic control by Poletti & Osta under general specification and system design of Centro Ricerche Fiat and with the IVECO design for the mechanical for the in wheel integration.

	D.C. SERIES EXCITED	D.C. SEPARATELY EXCITED	A.C. ASYNCRONOUS	D.C. BRUSHLESS WITH PERMAN. MAGNETS
FEATURES	* WIDELY USED FOR TRACTION (E.G. FORK LIFTS, INDUSTRIAL TRUCKS)	* EASY CONTROLLABILITY * POSSIBILITY OF CONTROL WITHOUT GEAR SHIFTING OR WITH GEAR SHFTING AND WITHOUT ARMATURE CHOPPER * HIGHER SYSTEM PERFORMANCE	* MAINTENANCE FREE * HIGHER SPECIFIC POWER	* MAINTENANCE FREE * HIGHER SPECIFIC POWER AND EFFICIENCY * LOSS FREE ROTOR * REDUCED ELECTRONICS VERSUS A.C. ASYNCRONOUS
PROBLEMS/ DEVELOPMENT NEEDS	* NEED OF GEAR SHIFTING TO MEET MAX. SPEED AND ACCELERATION REQUIREMENTS	* SLIGHTLY HIGHER COST	* MORE ELECTRONICS REQUIRED * NEED OF SPECIFIC DEVELOPMENT	* COST OF PERMANENT MAGNETS * NE D OF SPECIFIC DEVELOPMENT

FIG. 18 - TRACTION MOTORS

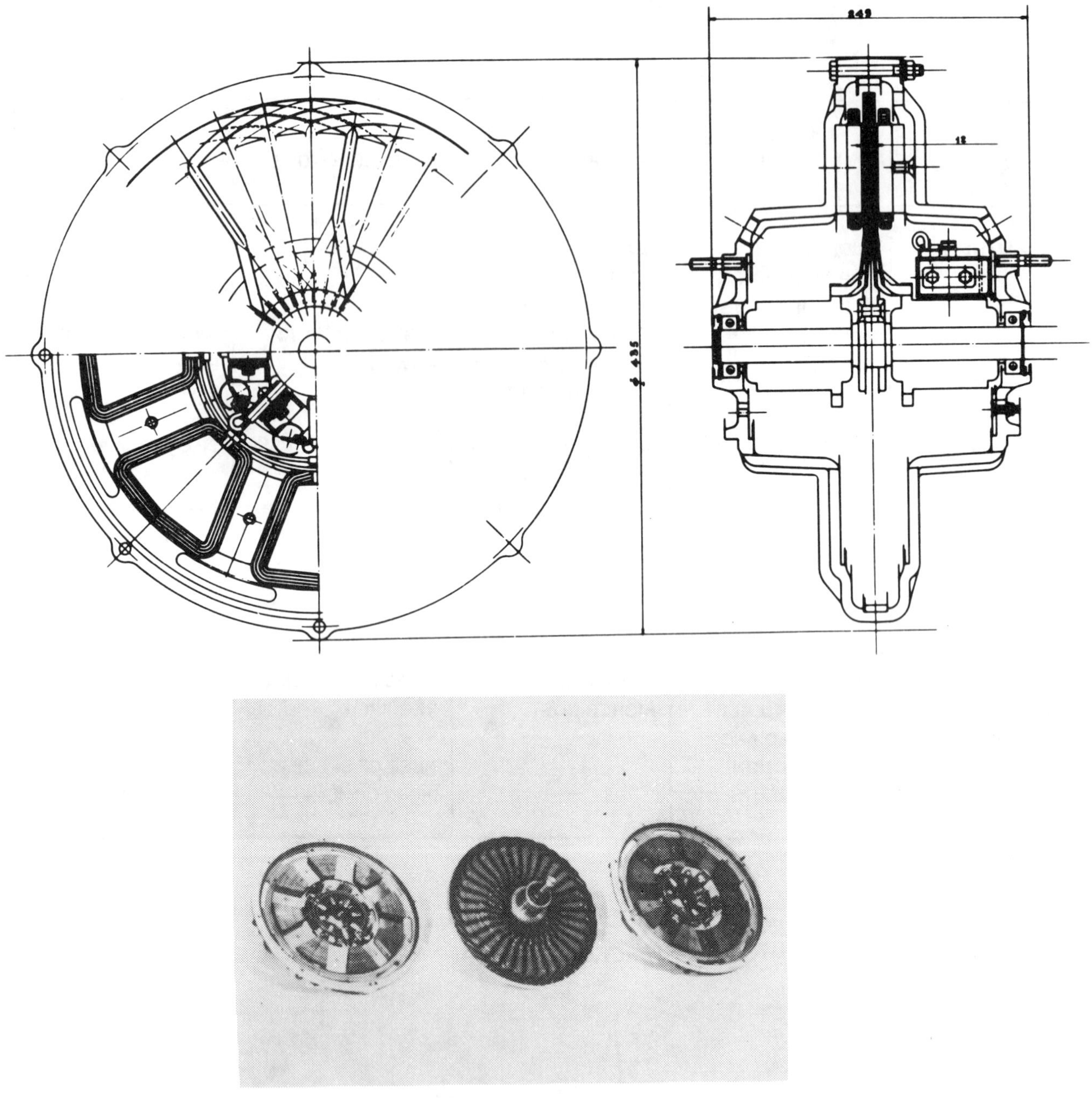

FIG. 19 - SECTION OF THE PERMANENT MAGNET D.C. MOTOR

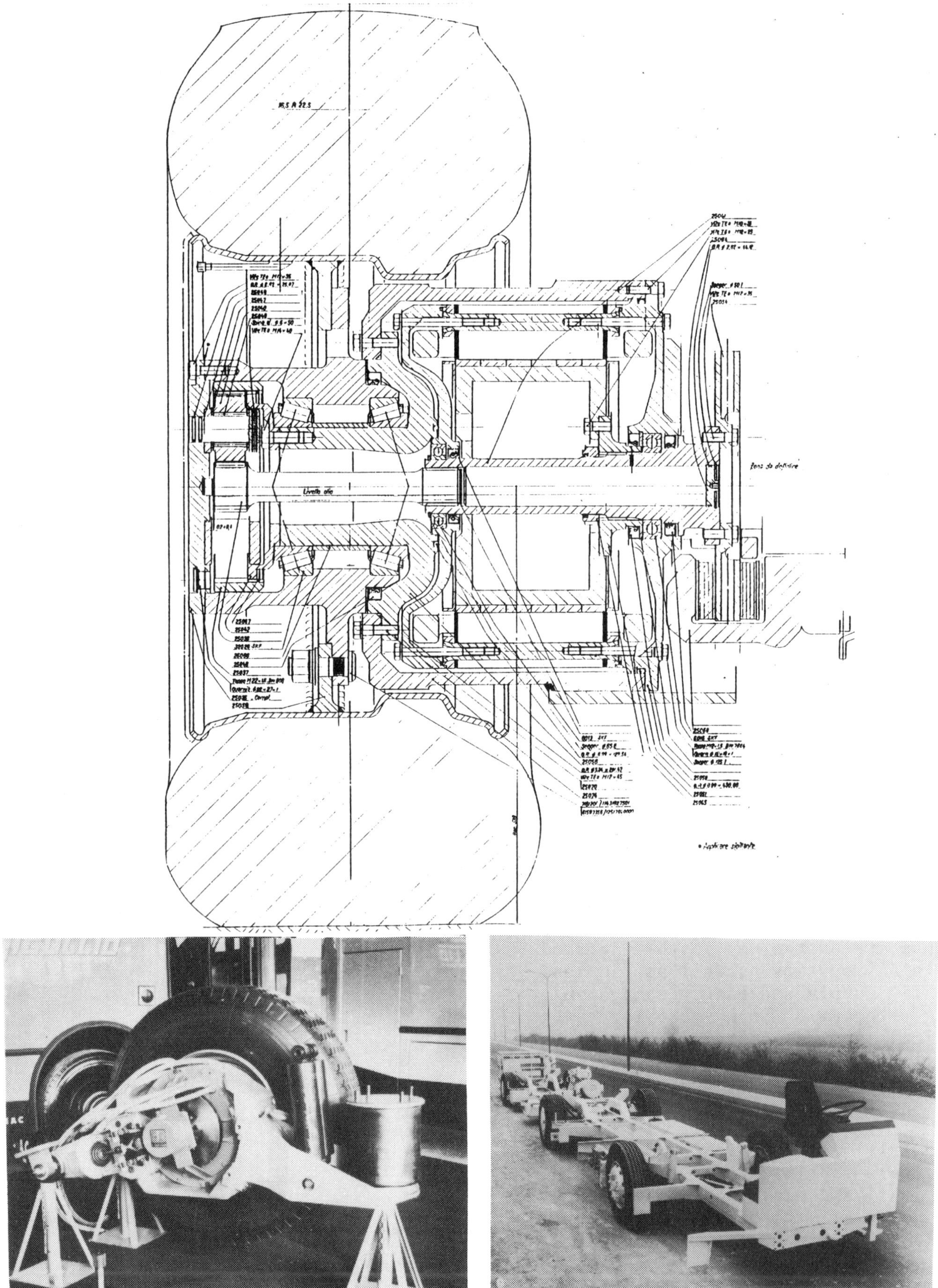

FIG. 20 - INTEGRATED MOTOR-WHEEL UNIT FOR MULTIMODE BUS

Four motor wheel units are applied to the second and third axle of an articulated 18 m multimode vehicle developed as follow-up of the previously mentioned 12 m multimode vehicle in the framework of the CNR Finalized Project Transport under-coordination of ATM of Milano (Fig. 20).

Another example of use of new technologies is given by the racing electric car Y10 which has been ad hoc developed from the model Y10 making use of Ni-Cd batteries and permanent magnet brushless motor.

CONCLUSION

The technological achievements experienced in many years of development allow practical use of electric vehicles for diversified utilization sectors, in pure electric solution, hybrid or multimode /8/.

The general concept followed in FIAT's development in this field is to offer electric vehicles with the same level of active and passive safety of the conventional vehicles, and with convenience, mobility and driveability features comparable with those of the today's urban vehicles.

While a production and utilization is going on with the validated technologies, new approaches and concepts are on the way in the field of propulsion systems as well as in that of energy supply and energy management, having the aim of improving the technical and economical figure of the electrically propelled vehicles of various kinds and contribute therefore to extend their application potential.

The ultimate solution requires the synergical optimization of all the mechanical and electrical parameters, which are at the basis of the finl vehicle performance.
Therefore, in parallel with the activity line dedicated to possible application of new energy sourcer, emphasis is given to research and development aimed at the best utilization of the on board energy; in this contact major attention is given to possible use of buffer units, reduction of mechanical losses and high efficiency propulsion systems .

Fig. 21 presents a scenario of possible introduction of new technologies and concepts for road electric propulsion.

SCENARIO OF INTRODUCTION OF TECHNOLOGICAL ACHIEVEMENTS FOR ELECTRIC VEHICLES

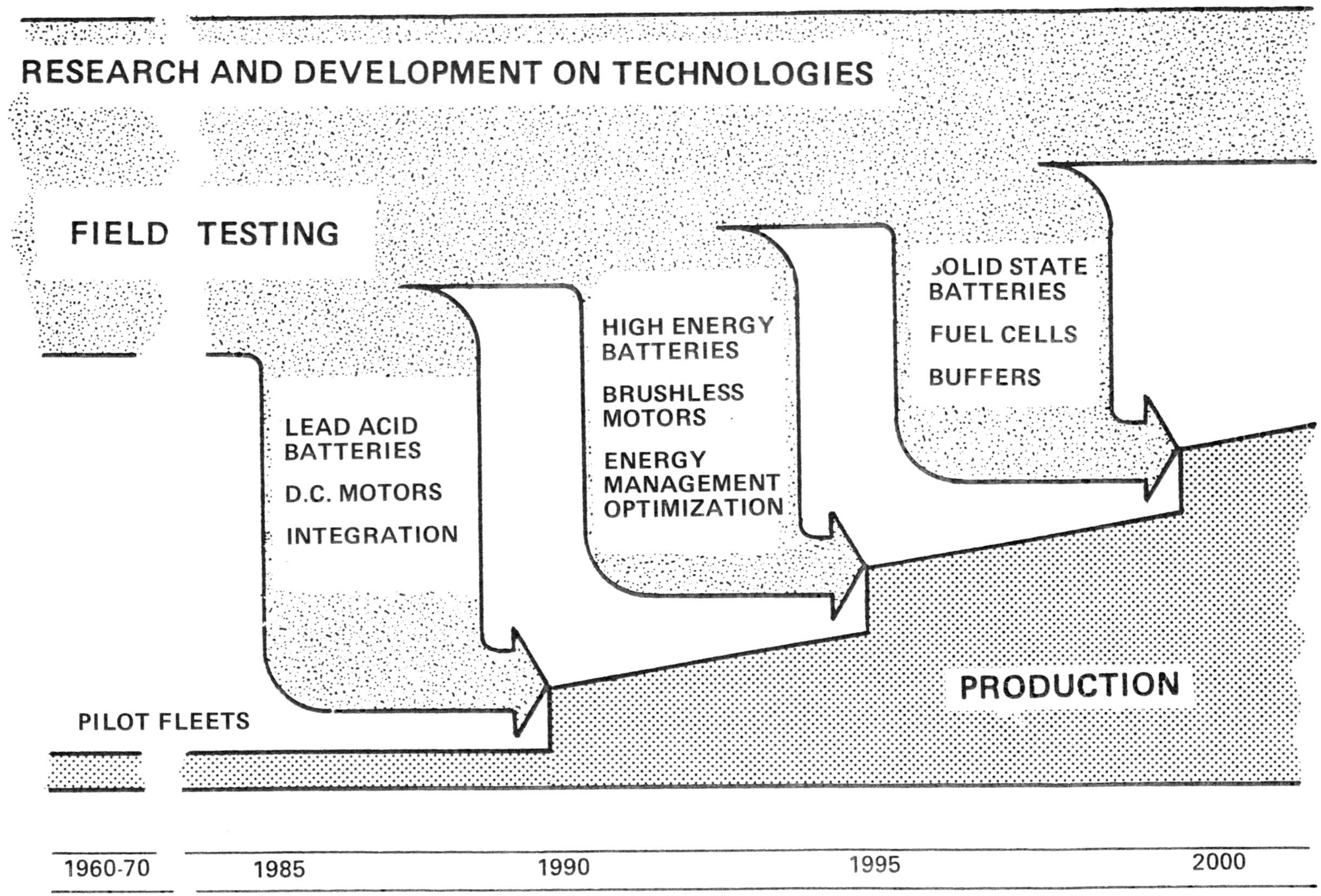

FIG. 21

REFERENCE

/1/ G. Brusaglino - Status and Prospect of Electric Vehicles in Europe - Proceedings from Tokyo EV Forum, September 1986

/2/ G. Brusaglino - The CNR Demonstration Program "Pilot City" in torino, Italy - Proceeding of EVC Expo '83 - Detroit, Michigan, U.S.A., October 1983

/3/ G. Brusaglino - Hybrid Electric Bus - Proceedings of the 12th meeting of Bus and Coach Experts - Budapest, June 1981

/4/ L. Mazzon, G. Brusaglino - Experience innovatriees et perspective d'evolution des vehicules de transport public sur pneus et rails - Proceedings of international Transport Congress - Montreal, September 1984

/5/ L. Mazzon - The role of electric vehicles in mass transport systems - Proceedings of the Conference DRIVE ELECTRIC ITALY '85 - Sorrento - October 1985

/6/ G. Brusaglino, G. Brocchetti, R. Gambarara, G. Polizzotto - THE CNR - FIAT IVECO 471 BM Trolleybus - Proceedings of the International Seminar C.E.C. - COST, Bruxelles 1985

/7/ Final Report of Action COST 303 - Docum. EUR 10993, 1987

/8/ G. Brusaglino - Electric propulsion development in Fiat for diversified applications - Proceedings of teh EVS-10 Symposium - Hong Kong, December 1990.

910245

Electrically Propelled Vehicles at BMW — Experience to Date and Development Trends

Hans-Hermann Braess and Karl-Nikolaus Regar
BMW

ABSTRACT

Back in the first two decades of automobile development, electric propulsion was a serious competitor for the internal combustion engine.

Electrically-propelled vehicles, however, soon proved unable to satisfy users' increasing performance demands in terms of range, acceleration, top speed and hill-climbing, together with such factors as operating life, initial purchase price, running costs and reliability.

Engineers investigating electric propulsion today face precisely the same unwelcome legacy as their predecessors, despite many and varied attempts in the meantime to improve the components of the electric vehicle's drive system (energy storage device, motors, controller).

Progress in battery development, particularly in the case of the NaS system, has nevertheless enabled us at least partly to overcome the previous problems associated with electric drive systems, though it cannot be claimed that all obstacles to its commercial application have been eliminated as yet.

This is a report on the experience gained at BMW to date, the current state of work and development trends.

CURRENT STATE OF DEVELOPMENT EVALUATED

Powered road-vehicle transport did not begin when the internal combustion engine was invented; this was predated by the battery-powered electric motor. The first electric vehicle with non-rechargeable batteries was built by Thomas Davenport in the USA as early as 1834, but at the same time first tests were taking place in Europe [1].

For reasons still fundamentally relevant today, it proved impossible to follow up these initial successes, with the result that electric propulsion later became restricted to a small number of special applications, notably local delivery vehicles.

In the course of time there have been numerous attempts on the part of inventors, state-run organizations and universities, but also from within the automobile industry, to revitalize the electric drive principle.

An early example of this is the BMW electric automobile which was used during the 1972 Olympic Games in Munich to accompany the marathon runners (Fig. 1). A further example is the "City Stromer", of which a small series was actually produced.

Range:	45 km
Battery weight:	350 kg
Engine output:	12 kW

Fig. 1: BMW electric vehicle with lead-acid batteries, 1972

There have been unceasing efforts on the part of research scientists to overcome the prime weakness of the electric drive principle, namely the device used to store the electrical energy.

In 1980 the company then known as Brown, Boveri & Cie (now ABB) successfully operated the first laboratory specimen of a novel type of high-energy battery, the sodium sulfur NaS) battery [2].

Since the specific energy density of this battery promised to represent a considerable step forward compared with the lead-acid battery, BMW resolved to recommence its research activities in the electric propulsion area. Initial feasibility studies were followed by several stages of development work [3, 4] which finally led to the current state of development as shown briefly in Fig. 2.

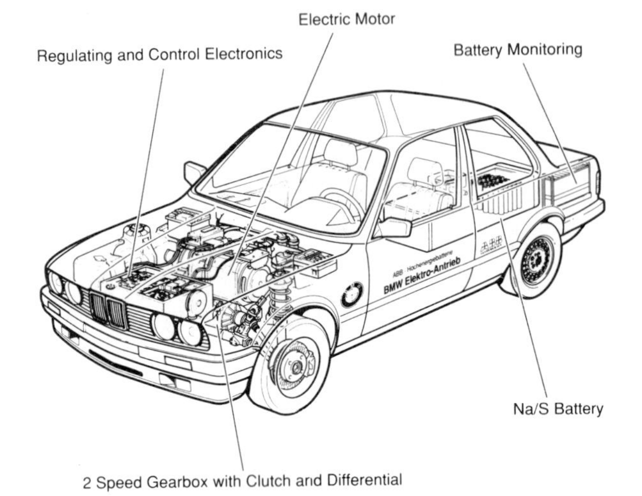

Range:	approx. 150 km	Engine output:	17/25 kW
Top speed:	100 km/h	Battery capability usable:	19 kWh
Acceleration:			
0-50 km/h	9.0 sec	Battery weight:	265 kg
0-80 km/h	23.0 sec	Battery volume:	248 dm^3

Fig. 2: BMW electric vehicle with NaS battery, 1990

As can be seen,

- the range and performance of this automobile can be regarded as adequate for town use, despite the fact that its design concept is not ideal for an electrically-propelled automobile
- the car's usability has been somewhat restricted in terms of luggage space and seating comfort, particularly at the rear, by the relatively large battery, but remains acceptable
- other criteria too, such as passive safety, convenience and day-to-day practicability, can be satisfied in full; however,
- the relatively large amount of space required for the battery has called for quite extensive modifications to the vehicle.

The following considerations speak against commercial utilization of the vehicle in its present form:

- the battery's lifetime is still not satisfactory
- this type of battery is still very much too expensive
- its power density is not yet satisfactory and the motor/controller unit operates at relatively poor part-load efficiency
- these components are currently still very expensive.

These are the very reasons why no major automobile manufacturer has so far marketed a model entirely adapted to electric propulsion.

However, the need to adopt electric automobiles continues to grow in an era in which it is becoming essential to achieve a further reduction in emissions in urban areas and in which every possible way of safeguarding the energy base needed for transport must be fully exploited.

The purpose behind the electric automobile must be not only to reduce pollutant emissions and CO_2 where it is operated, but to contribute towards the same effect in general.

Fig. 3 is a comparison in this respect between a BMW automobile with spark-ignition engine and closed-loop, three-way catalytic converter, and a derivation of the same model with electric drive.

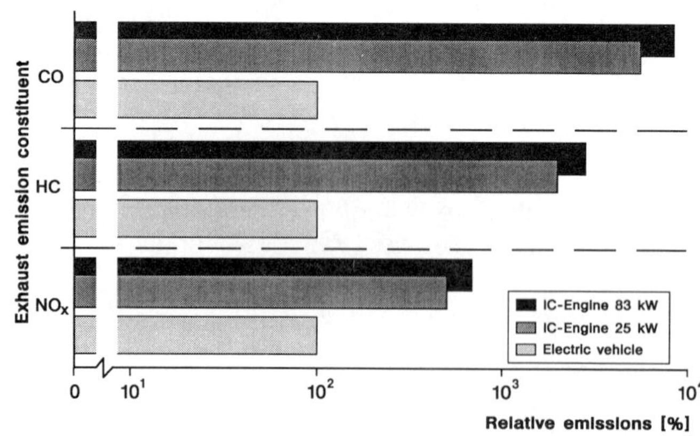

Fig. 3: Relative emission patterns from electric and spark-ignition engined vehicles (including extraction/generation of primary energy)

These values were determined using the BMW cars' measured energy consumption and the specific emissions resulting from electricity generation, taken from a publication issued by the RWE, a leading German electricity generating company [5]. The relationship applies to the current electricity generating mix applicable to the Federal Republic of Germany.

In the case of CO_2 emissions the situation is not entirely clear, since the specific energy consumption of electric vehicles with an NaS battery also depends on their operating profile: the greater the distance covered daily, the more favorable the specific energy consumption and the associated CO_2 emissions from the electricity generation process. The cause of this is to be sought in the thermal losses from this type of battery.

Fig. 4 is a comparison between BMW vehicles.

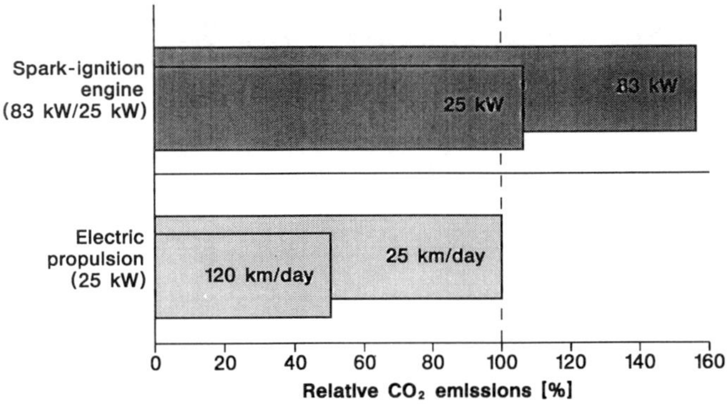

Fig. 4: Relative CO_2 emissions from electric and spark-ignition engined vehicles (including generation/extraction of primary energy)

The values shown for electric automobiles apply to a distance of approx. 27 km/day (260 Wh/km), with no weekend operation [6], and to a daily distance of 120 km (130 Wh/km), as measured for instance during the 1990 AUSTRO SOLAR RALLY.

The fuel consumption for vehicles with internal combustion engine is the average of the three standard European test values.

Such results clearly indicate that despite all the currently unsolved development problems the electric automobile will in future be capable of playing its part as an individual mode of transport in urban areas.

Some of these problems will prove easier to solve if ingenuity is applied to questions of system component selection, rating and mode of operation (batteries, motors, additional storage devices, drive system management).

The experience gained by BMW in this respect is described below.

ELECTRIC VEHICLE REQUIREMENTS AND THEIR DEPENDENCE ON THE DRIVE SYSTEM

Fig. 5 shows that the electric vehicle's ability to satisfy most of the major requirements imposed on it depends very greatly on the drive system and the characteristics of its components.

Requirement:		Dependent on:
- Range	▲	Energy storage device
- Acceleration, hill-climbing and adequate steady speed	▲	Nominal battery resistance motor torque motor power output
- Economical use of energy	▲	Drive system efficiency
Low weight combined with economical use of space		Energy- and power density of battery
- Life time of energy storage device	▲	Operating conditions
- Ergonomically efficient control and refinement	▲	Vehicle design ratings
- Pleasant interior climate	▲	Design and rating of heating and ventilation system
- Passive safety	▲	Overall vehicle safety design concept
- Economy	▲	Initial purchase cost, operating costs

Fig. 5: Basic requirements for electrical vehicles and their drive systems

BATTERIES - Fig. 6 summarizes the principal requirements which modern traction batteries must satisfy. In practice, many of these demands amount to a conflict of objectives, since on account of their operating principles none of the battery systems known to be suitable for traction purposes satisfy them all equally well.

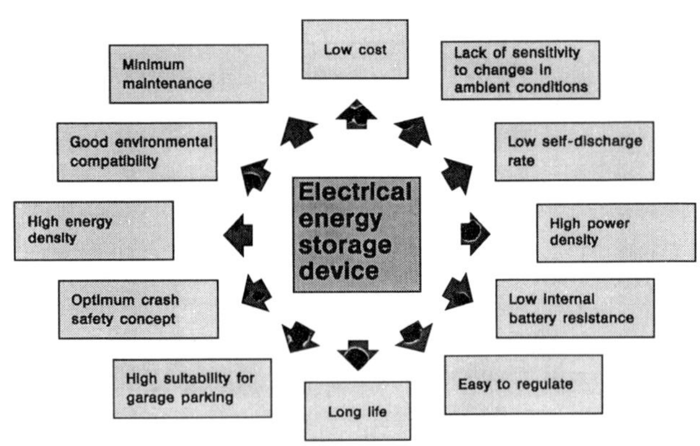

Fig. 6: Important requirements for electrical energy storage devices in electric vehicles

With reference to the basic requirements, namely power density, energy density, purchase price and life expectancy, Fig. 7 shows which conflicts of objective have to be taken into account by the motor-vehicle engineer when selecting batteries for an electrically propelled automobile.

Battery system	Energy density	Power density	Running life	Manu-facturing costs
Lead-acid	Low	Good	Low	Acceptable
NiCd	Adequate	Very good	Adequate	Very high
$NaNiCl_2$	Good	Adequate	Low	Very high
NaS	Very good	Good	Low	High

Fig. 7: Important assessment criteria for traction batteries

ELECTRIC MOTORS AND CONTROLLERS - A similar situation is found to apply in the case of electric motors and controllers.

Figs. 8 and 9 show the criteria which these items of equipment must satisfy and the resulting design conflicts.

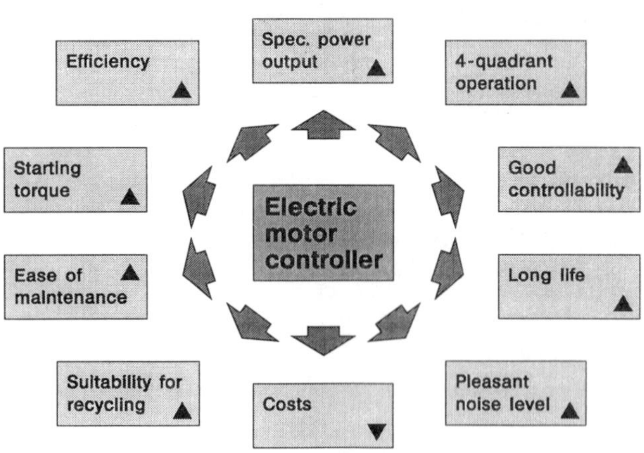

Fig. 8: Important requirements for electric vehicle motor-controllers

Motor-Controller concepts	Requirements			
	Spec. power output	Starting torque	Efficiency	Costs
DC motor	Low	Adequate	Unsatisfactory	Unsatisfactory
Threephase motor	Adequate	Adequate	Satisfactory	Unsatisfactory
Permanent-magnet motor	Very high	Low	Good	Unsatisfactory
DC motor with electronic commutation and integral controller	High	High	Very good	Satisfactory

Fig. 9: Important assessment criteria for electric vehicle motor- controller concepts

None of the motor-controller concepts currently offered is capable of satisfying requirements to a sufficient degree. Efficiency in the lower part-load range in particular needs to be greatly improved.

Since there did not seem to be any great willingness on the part of the supplying industry to eliminate this shortcoming, BMW Research tackled the development of a suitable motor itself.

Its principal characteristics are:

- high power density (approx. 1 kW/kg including controller)
- high specific torque (approx. 8 Nm/kg)
- excellent efficiency over the entire operating characteristic including the lower part-load range
- potentially low manufacturing costs.

Current work is being devoted to improving the strength and reliability of the mechanical and electronic systems.

CONVERSION OF OUTPUT CHARACTERISTIC (TRANSMISSION) - At a given rolling resistance for the vehicle, performance is not only determined by the motor (as it is by the engine of an internal combustion engined vehicle) but by the power available from the energy storage device.

This relatively low amount of electrical power therefore has to be converted into mechanical energy at the highest possible rate of efficiency and transmitted in the same way to the wheels, if necessary by means of a gearbox and differential.

The spread of gear ratios to be provided by the transmission is determined by the motor speed, final drive ratio and the road wheel radius. The same applies to the maximum gradient which the vehicle is required to surmount.

In view of these determining factors, and allowing for the fact that the electric automobile is intended for urban use, it becomes possible to define the transmission concept, as shown in Fig. 10.

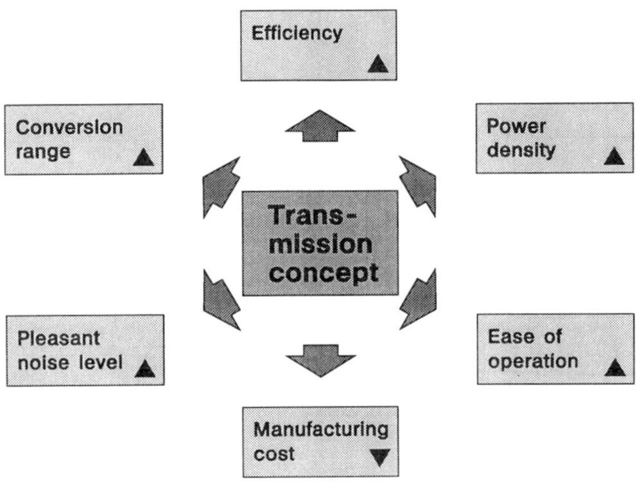

Fig. 10: Important requirements for transmission concepts in electric vehicles

In this area too, vehicle development staff have had to depart from familiar techniques, since merely adopting an existing transmission concept (multi-ratio manual shift gearbox, automatic or CVT transmission or even fixed drive ratios) does not normally yield the required results in conjunction with the electric traction motor (see Fig. 11).

Trans-mission concepts	Requirements				
	Conversion range	Efficiency	Power density	Easy of Operation	Costs
Manual-shift gearbox	Good	Moderate	Good	Poor	Satisfactory
Automatic transmission	Good	Poor	Poor	Adequate	High
CVT	Moderate	Poor	Very poor	Good	High
IVT	Very good	Moderate	Poor	Good	Very high
Fixed-ratio transmission	Inadequate	Very good	Very good	Very good	Very low
2-speed manual-shift gearbox	Adequate	Good	Good	Very good	Low

Fig. 11: Important assessment criteria for electric vehicle transmission concepts

DRIVE SYSTEM CONCEPTS FOR ELECTRIC VEHICLES

The inadequate state of development and the presence of conflicts of objectives as far as the drive system components - batteries, motors and transmissions - are concerned have confronted motor vehicle engineers with unconventional tasks. This is borne out by some misdirected lines of development and unexpected setbacks which have occurred in this field: although the connection of a battery to a motor and the output of power through a transmission may seem straightforward enough, a satisfactory solution seldom results in practice.

Aware of these problems, BMW has from the outset endeavored to make optimum use of the strong points of the NaS battery (for example its high energy density) and to compensate for its weaknesses (for example relatively low power density), as they became manifest from 1981 onwards.

By combining this battery with a mechanical storage device (flywheel or gyro), it became evident that a sensible division of functions could be achieved:

- the NaS battery's high energy density would enable the desired range of 100 km or more to be achieved with a relatively small, light battery, whereas
- the gyro's high power density could be used to obtain the desired rates of acceleration.

Systems of this type already tested and built in conjunction with lead-acid batteries [7, 8] were found to be unsatisfactory for a variety of reasons, and therefore recourse was had to the principles shown in [9]. This concept was thoroughly revised to suit the specific application, leading to the drive system shown in schematic form in Fig. 12.

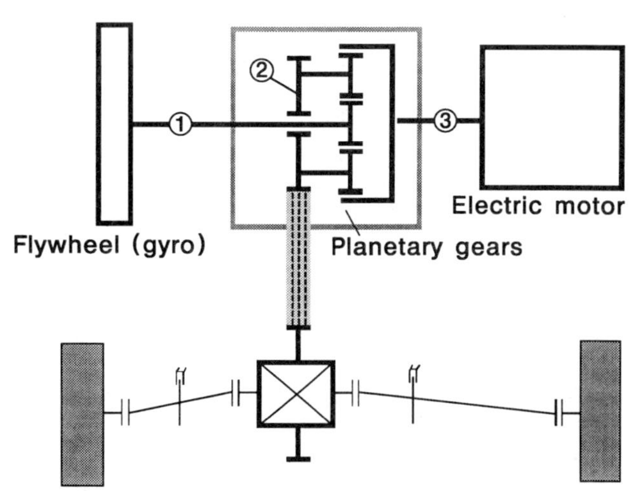

Fig. 12: Basic layout of an electric drive system with flywheel (gyro)

The three shafts of a planetary gearbox are connected as follows:

- Shaft 1 (sunwheel) to the gyro
- Shaft 3 (annulus) to the electric motor, and
- Shaft 2 (spider) to the driven axle.

Planetary gearbox, electrical machine and battery (not illustrated here) together form a transmission with an infinite adjustment range. The infinitely variable transmission (IVT) is familiar to us from published references and can also be obtained using hydrostatic machines and high-pressure reservoirs in place of electrical machines and batteries [10, 11].

Such transmissions permit energy to be transmitted steplessly from the gyro to the driven wheels or vice versa, as a result of which the energy generated by braking the vehicle can be stored in the gyro.

The power flow is determined by the speed of rotation and torque of the electrical machine. Both values (speed and torque) can be continuously varied over a wide range by means of signals from the accelerator or brake pedals and from a road speed sensor.

The power requirement at the wheel and the gyro's speed of rotation are given values between which a fixed ratio exists as a result of the static gear ratio of the planetary gearbox.

The electrical machine and the battery thus form the variable path of the infinitely variable transmission, and accordingly only a small proportion (approx. 30 %) of the energy flow is allowed to pass along this path, the larger part (approx. 70 %) being transmitted entirely mechanically at high efficiency by way of the sunwheel (shaft 1) and spider (shaft 2).

It would of course also be possible to duplicate the function of this infinitely variable transmission with the aid of two electrical machines, one coupled to the flywheel, the other to the vehicle's driven axle. However, this arrangement has a whole series of disadvantages compared with the combined mechanical and electrical approach:

- To obtain the same power at the wheel with electrical machines alone calls for these to possess at least twice the total power of the machine as shown in Fig. 12, since there is no mechanical power addition by way of the planetary gears.
- electrical machines of twice the output power would mean distinctly higher weight and increased costs and would occupy more space.
- compared with the current state of the art, energy transmission efficiency would be unacceptably poor.

The electric/gyro drive system shown schematically in Fig. 12 is only capable of realizing transient operating conditions: charging the gyro energy store, accelerating and braking.

In order to permit lengthy distances to be covered at a constant or almost constant road speed, it must be possible to decouple the gyro from the drive system.

Furthermore, a reduction gear stage is needed to match the speed of rotation of the gyro energy reservoir to the speeds at which the electric motor is permitted to run and to the vehicle's road speed.

The complete drive system is shown in Fig. 13.

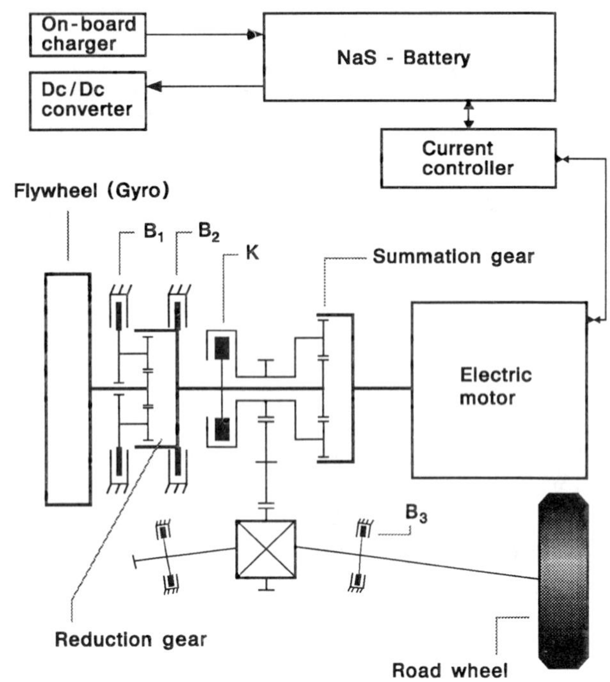

Fig. 13: Drive system with flywheel energy storage device for electric vehicles

Its mode of operation is as follows:

- Brakes B_1 and B_3 are engaged before starting the journey, to charge the gyro storage device. The electrical machine is supplied with current from the battery and operates as a motor. It rotates in a negative direction, the gyro output in a positive direction.
  At the end of the charging phase, the flywheel is rotating at its maximum speed (in this case 14,000 min^{-1}), and the electrical machine revolves off-load in the negative direction at a speed determined by the ratio in the reduction gears and the summation gears.
- When acceleration of the vehicle is required, the electrical machine is switched to the generator operating mode and thus braked to a standstill.

At this operating point, the electrical machine's output (though not its torque) is zero and the gyro storage device's energy is transmitted entirely mechanically to the wheels. From this point onwards, however, the electrical machine must be switched to the motor operating mode (with current from the battery), after which its speed increases in the positive sense of rotation until acceleration has been completed. The system is designed such that the end of acceleration coincides with maximum road speed, the desired minimum gyro speed and the maximum speed of rotation of the electrical machine (see Fig. 14).
In the lower road-speed range (= less power required in a steady state), up to the point at which the electrical machine reverses, the system is supplied exclusively with energy from the gyro storage device; in the upper range (= greater power required) it obtains its energy as appropriate from the gyro storage device and the battery.
This combination provides almost twice the acceleration power at the wheel in certain circumstances, referred to the output power of the electrical machine.

- When braking, these operating phases take place in the reverse sequence, with similar effect.
- As soon as acceleration has ended, the gyro is disconnected (brake B_1 open, clutch K closed) and the vehicle is driven by the electrical machine alone.
- If acceleration ceases at a fairly low road speed (below 50 km/h), a higher ratio is selected between the electrical machine and the differential gears, by leaving clutch K and brake B_1 open and applying brake B_2.
- The system can therefore also be run with the gyro storage device disconnected from the drive system, and the planetary summation gears operating as a 2-speed automatic transmission.

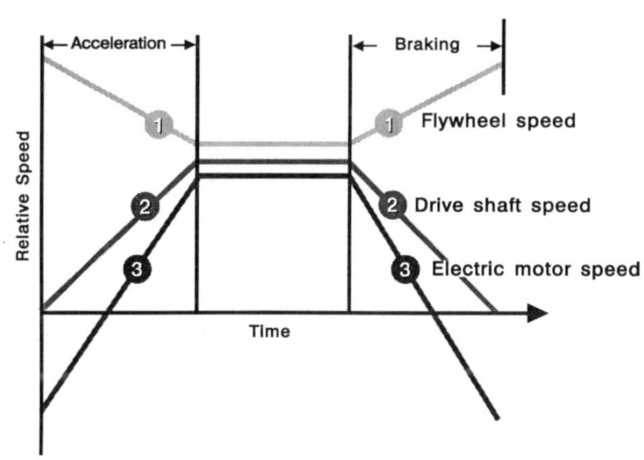

Fig. 14: Relative speeds of rotating drive components during a driving cycle

Fig. 15 shows measured acceleration times from 0 to 80 km/h when running the vehicle with and without the gyro storage device.

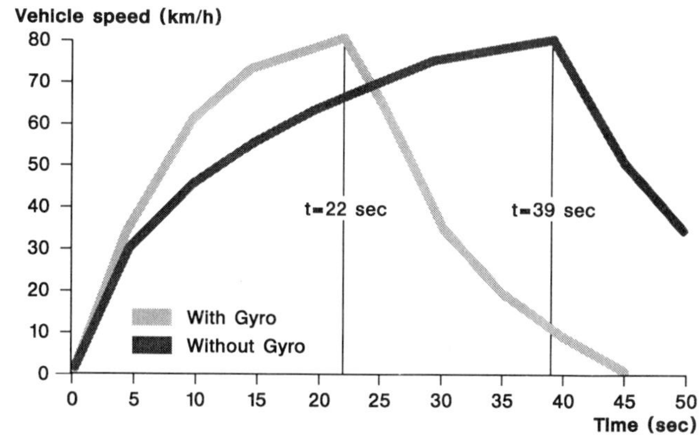

Fig. 15: Acceleration potential of electric vehicle with and without gyro

Fig. 16 illustrates another effect which could be of decisive importance with certain types of battery (for example the lead-acid battery): despite the high acceleration performance obtained with the gyro in operation, a distinctly lower battery load with high peak currents is incurred.

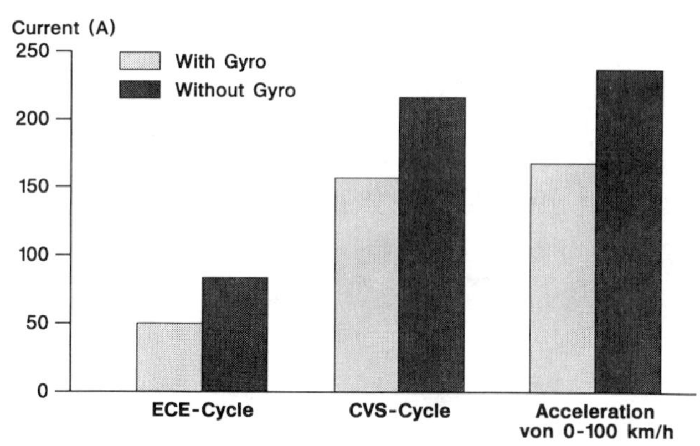

Fig. 16: Variations in load on battery when operating with and without gyro

Lead-acid batteries can therefore achieve an extended operating life and higher efficiency.
Fig. 17 shows the result of a specific driving cycle: the vehicle was accelerated repeatedly from 0 to 50 km/h, then braked immediately to a standstill, the cycle being repeated after an interval of 6 seconds. It can be seen that the reduced load on the battery when operating with the gyro has a highly positive

influence on the uniformity of the acceleration results throughout the series of driving cycles. In other words, the "battery fatigue" from which the lead-acid battery suffers is less pronounced: the acceleration times in the 1st and 15th cycles increased only slightly with the gyro in operation (from 16 to 19 seconds), but became significantly longer (from 16 to 30 seconds) without the gyro.

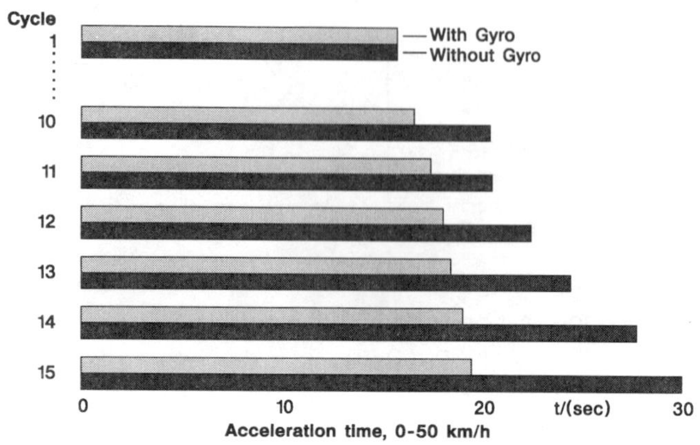

Fig. 17: Variation in acceleration times during cyclic operation with and without gyro

Fig. 18 shows the experimental drive system.

Fig. 18: Electric drive system with integrated gyro energy storage device

The vacuum pump needed to evacuate the gyro housing and above all the oil pump which actuates brakes B_1 and B_2 and clutch K have proved to have an adverse effect on the efficiency of the system as a whole. The pumps which we tested were of very low efficiency at the required pressure (5.10^5 Pa) and delivery rating (approx. 5 l/min).

The same applies to the electric motor which drives this pump. At low outside temperatures its power consumption while the oil was still cold was up to three times higher than when operating at normal oil temperature.
In order to minimize friction losses and keep the weight of the flywheel down, a maximum energy storage rating of 125 Wh at 14,000 min^{-1} was chosen; of this, approx. 100 Wh were usable energy. This energy capacity was sufficient for one full-load acceleration or braking operation.
The loss rating of the gyro with reduction gears was 1.9 kW at maximum speed.
This relatively low figure applies at an oil temperature of 340 to 360 K and a vacuum of $0.35.10^5$ Pa (0.35 bar) in the gyro housing. But even at lower oil temperatures (T = 300 to 310 K), the loss rating does not exceed 2.2 kW.
The complete assembly, comprising electric motor, summation gear, reduction gears, gyro energy storage device, differential, vacuum pump and oil pump, weighed 238 kg, of which the electric motor alone accounts for 90 kg.
The measured results did not yield a clear picture as far as energy consumption was concerned. In the case of driving cycles with a lower dynamic factor (ECE test cycle), consumption was lower when operating without the gyro, whereas on the FTP-75 test cycle consumption was lower with the gyro. However, we can assume that further improvement to the system, for example elimination of the reduction gears, will give the gyro energy storage concept a clear advantage in energy terms.

These results were obtained when running the vehicle on lead-acid batteries.

A change to the NaS battery revealed that the ceramic electrolyte of this battery reacts with great sensitivity to overvoltages. For this reason the electrical machine's power output in the generating mode (see Fig. 14) had to be severely restricted.

The consequence of this was that during braking, when the generating mode predominates in terms of time and in particular in terms of energy flow, the braking power was far too low, with the result that only a small proportion of the braking energy occurring at the road wheel could be supplied to the gyro storage device. Renewed full-power acceleration was then possible only after the gyro had been recharged with energy taken from the battery.

This situation on the one hand greatly lengthened the intervals needed between acceleration phases and on the other hand significantly reduced the efficiency of the entire system. There were also certain technical problems related to control of the switching process between transient and steady-state operation and vice versa.

In other words, the advantages of the gyro energy storage device, although clearly evident in conjunction with lead-acid batteries, were distinctly lower when used with the NaS battery.

However, since ABB changed the ratio of energy density to power density from 1.18 in 1981 to 0.9 by 1986, the battery capacity chosen for the desired range is now also sufficient to ensure adequate acceleration (see Fig. 2).

This enabled us to dispense with the gyro as an aid to acceleration; current experimental vehicles perform adequately with a simple 2-speed manual-shift gearbox (Fig. 19).

Fig. 19: Electric drive unit for current experimental vehicles

An NaS battery with high storage capacity and relatively low power output is only practicable without an aid to acceleration if the electric car's range is at least 100 kilometers, or else insufficient power will normally be available to achieve acceptable acceleration.

However, if only a limited range (for instance 50 km) can be accepted, and for cost reasons the batteries are to be designed for this distance, it becomes almost essential to combine this battery with a gyro storage device.

For this reason the drive system described here, despite its disadvantages, remains an interesting and practicable proposition.

By means of new developments in motor technology it will prove possible to eliminate the drawbacks of the system investigated here completely.

IMPORTANT TASKS FOR THE FUTURE

Despite the advanced state of development already attained, electrically powered vehicles are not yet suitable for commercial production.

The price of the battery in particular is still too high, and its life expectancy too low.

Intensive work will be needed to improve these situations.

Current knowledge suggests that development work should not be concentrated on a single battery system. As well as the NaS battery, which undoubtedly represents the most advanced current high-energy battery concept, NiCd, $NaNiCl_2$ etc. still offer worthwhile prospects of success.

Electric motors need a great deal of improvement, but certainly possess the necessary improvement potential.

Integration of the power and control electronics into the design of the motor helps, for instance, to keep overall dimensions compact and weights low, improves efficiency, avoids the need for mechanical commutation and achieves a drastic improvement in the electromagnetic compatibility of the drive system compared with current motors equipped with a separate controller.

Vehicle designers too are confronted with a multitude of tasks, from the achievement of optimum drive system design through to optimum integration of the electric drive system into suitable vehicle concepts.

For example, the following tasks will have to be tackled as a matter of urgency in the near future:

- Improved braking power assistance
- Development of electrically power-assisted steering
- Electric heating
- Electrically driven air conditioning
- Improved thermal insulation of vehicle
- The complete vehicle's crash behavior
- Development of more efficient on-board chargers (approx. 97 % instead of the current 90 %)
- Low-loss supply of auxiliary consumers' energy requirements from a DC/DC converter

Further tasks will arise in the marketing, sales, after-sales service and waste material disposal areas when a market launch becomes imminent, and also in the energy supply and energy marketing areas.

Last but not least, our politicians will be called upon to create the appropriate general conditions for operation of these vehicles.

REFERENCES

[1] Büberl, A.
"Die Entwicklung der Elektrofahrzeuge bis zur Gegenwart" ("Development of Electric Vehicles to the Present Day"); in a special supplement to "Energie und Umwelt Aktuell", No. 3/1988

[2] Fischer, W. and Baukal, A.
"Der Natrium/Schwefel-Akkumulator als Antriebsquelle für Elektrofahrzeuge" ("The Sodium-Sulfur Battery as a Propulsion Source for Electric Vehicles"); in etz-a, 58 (1977), 75

[3] Regar, K.N.
"Stand der Entwicklung und Perspektiven des BMW-Elektroautos" (State of Development and Prospects for the BMW Electric Automobile"); lecture held in Bonn-Bad Godesberg, Germany, in December 1987

[4] Braess, H.H, Regar, K.-N. und Strobl, W.
"Drive Systems to Protect the Environment"; in Automotive Technology International 1990, publ. Sterling Publications International, London.

[5] Sporckmann, B.
"Elektrofahrzeuge als Luftschadstoffbremse?" ("Electric Vehicles as a Brake on Air Pollution?"); in "Energiewirtschaftliche Tagesfragen", No. 6/1990

[6] Blumenstock, K.U.
"Watt ihr Volt, BMW: Elektroauto kommt voran" ("Watt do you Want, BMW's Electric Car is Making Progress"); in "mot-technik", 26/1989

[7] Locker, D. and Miller, M.L.
"Flywheel-Electric Hybrid Vehicle - 4th International Electric Vehicle Symposium" held on 31. Aug./2 Sep. 1976 in Düsseldorf

[8] Proche, G.G.
"Elektroantrieb mit Schwungradunterstützung" ("Electric Drive with Flywheel Assistance"); in "Automobil Revue No. 4,
22. Jan. 1981

[9] Regar, K.N.
"Elektrische Antriebsbaugruppe mit mechanischem Energiespeicher für Strassenfahrzeuge" ("Electric Drive Assembly with Mechanical Energy Store for Road Vehicles"); unpublished development report, Apr. 15, 1976

[10] Regar, K.N.
"Stepless Transmissions with Hydrostatic Power Branch for Energy-saving Traction Drives"; Warwick, 1981, Paper C 153/81

[11] Regar, K.N.
"Untersuchungen zur Optimierung der Betriebsparameter eines hybriden Busantriebs zwecks Minimierung des Kraftstoffverbrauchs" ("Investigating Optimization of the Operating Parameters of a Hybrid Bus Drive in order to Minimize Fuel Consumption"); in Fortschritt-Berichte VDI-Z, Series 12, No. 43/1982

910246

Integrated Electric Vehicle Drive

William M. Anderson and Craig S. Cambier
Unique Mobility, Inc.
Englewood, CO

ABSTRACT

An advanced drive system for electric vehicles is described. The extremely compact and lightweight design features a high efficiency, high power density permanent magnet brushless dc motor with an integral gear reduction and differential. The electronic controller features full four quadrant operation with regeneration capable to zero speed.

INTRODUCTION

Unique Mobility is an Englewood, Colorado based R & D company active in the development of drive systems for electric and hybrid electric vehicles. The key technology is based on lightweight permanent magnet motors coupled with solid state electronics and forms the basis for the company's effort to develop an advanced electric drivetrain, that is capable of operating from any available source of electricity such as batteries, on board engine generators or fuel cells.

Unique's motor and controller technologies are currently being evaluated in various applications such as the recent GM sponsored Sunrayce from Florida to Detroit, Michigan, held in July 1990. Of the 32 university teams entered, 18 vehicles employed Unique motors and controls.

MOTOR AND CONTROLLER TECHNOLOGY

Figure 1 shows an exploded view of a typical Unique motor. Radially positioned permanent magnets are mounted on a hollow rotor which is co-axial with the radially thin stator winding. The stator windings are directly pierced by the magnetic field and a high magnet pole count is employed. The use of narrow magnets permits use of a relatively thin steel mounting ring for the rotor and a correspondingly thin stationary return path in direct contact with the stator winding. Figure 2 shows the stator winding in various stages of construction. No laminations are employed. Hysterisis and eddy current iron losses are minimized in this design. Copper utilization is also high as the ratio of active length to end turn length is much higher than in a conventional motor wound using silicon steel laminations. Continuous power densities on the order of 3 kW/kg are readily obtained using air or liquid cooling means. The winding structure has good mechanical strength because it is encapsulated in high temperature epoxy.

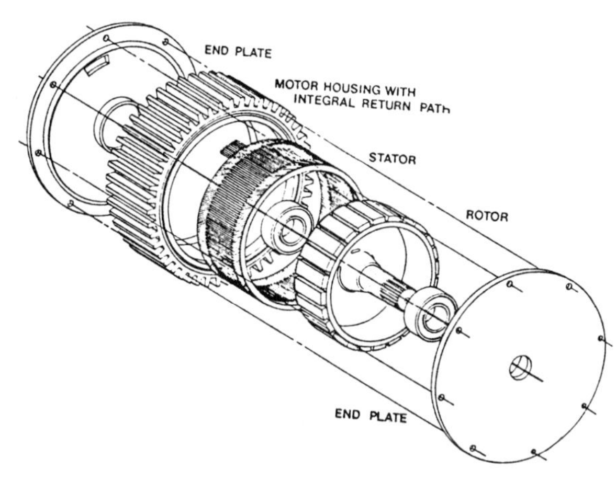

Figure 1

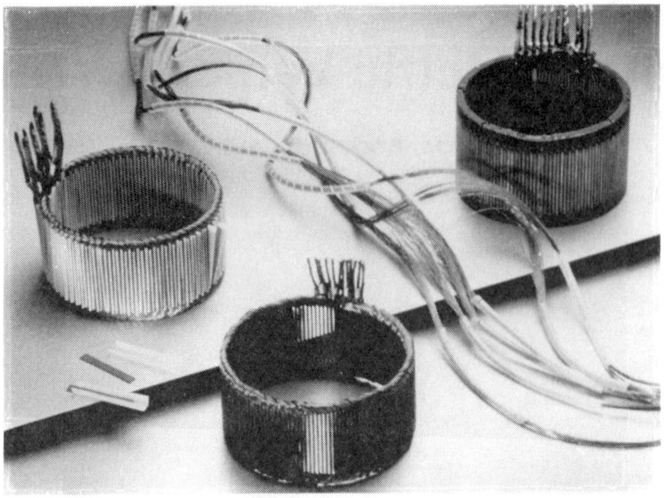

Figure 2

The motor has no brushes and is commutated electronically by using solid state devices. The power switching devices switch the current on and off to the motor windings in a series of pulses which controls the motor speed and torque. The characteristics of the motor are such that it has a very low inductance as well as a low resistance. High commutation frequencies (up to 1.5 khz) are employed. Both MOSFETs (metal oxides semi-conductor field effect transistors) and IGBTs (insulated gate bi-polar transistors) may be employed as they have the requiste switching speed. Figure 3 shows the basic controller topology typically employed for the 3-phase wye winding configuration of the motor.

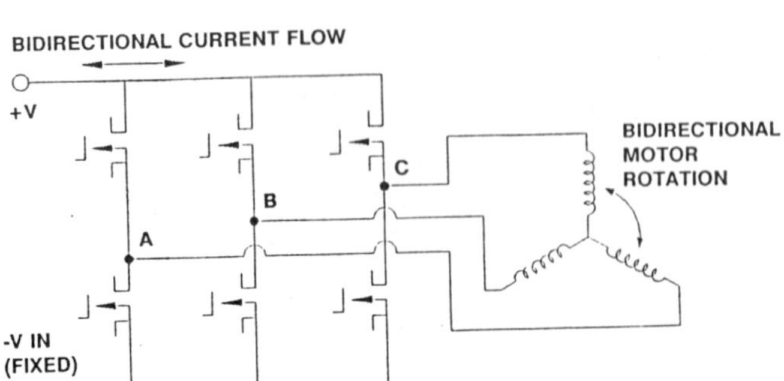

Figure 3

The controller is capable of full four quadrant operation. Regeneration is possible down to near zero speed. Hall effect sensors in the motor provide the timing signals required for commutation. Figure 4 shows a typical motor current wave form under pwm operation. The controller operates with an efficiency greater than 95% over most of the operating speed range.

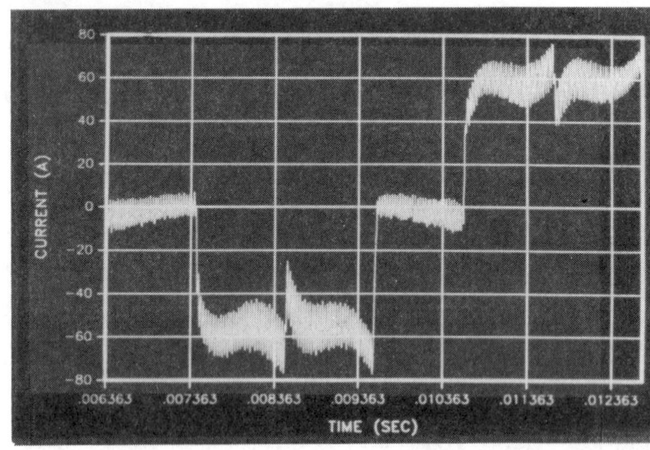

Figure 4

Figure 5 shows a nominal 50 kW motor and controller designed for application to an electric vehicle drive system. The motor weighs approximately 15 kg and is air cooled. The controller also air cooled and weighs 20.5 kg. Two of these motors are being used to power a hybrid electric minivan conversion currently being developed.

Figure 5

VEHICLE REQUIREMENTS

Typical road vehicle wheel torque versus speed characteristics are shown in Figure 6. The peak torque corresponds to that required for vehicle acceleration from a stop or low speed. The constant power region corresponds to acceleration requirements for merging and/or passing. The primary variable that defines the envelope of operation as shown in Figure 6 is vehicle weight. Values of peak wheel torque required are on the order of 500-600 Nm/1000 kg vehicle weight. In the region of constant power, 15 to 20 kW/1000 kg is typical. The power required for high speed operation (80-120 km/hr) is dependent primarily on aerodynamic drag and rolling resistance and is on the order of one-half of the constant power value.

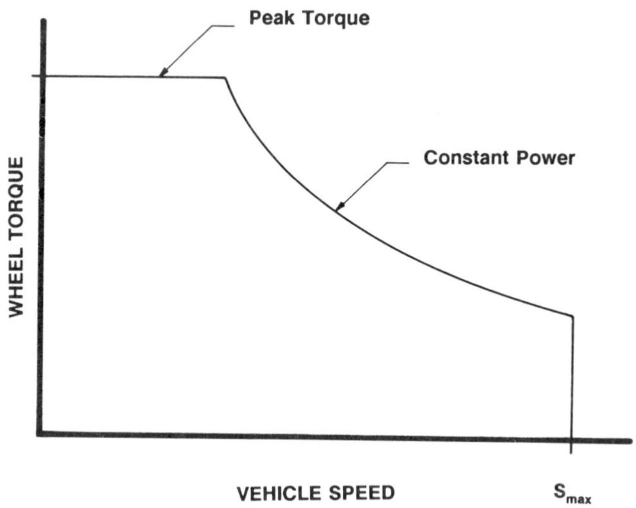

Figure 6

For a given vehicle, the engine or electric motor torque desired to achieve the required wheel torque depends upon the gear reduction between the prime mover and the wheels. The torque output versus speed curve of an internal combustion engine is relatively flat and with a single value of gear reduction cannot match the range of wheel torque requirements. Hence the need for multi-speed transmissions. For electric vehicles it is possible to consider a single ratio gear reduction, thus saving weight and increasing efficiency by eliminating the multi-speed transmission. Series or shunt wound brush commutated dc motors generally exhibit the desired torque versus speed characteristics and many successful electric vehicles have been built using such motors. However, these motors are quite heavy and are less inefficient than permanent magnet brushless dc motors or induction motors operating at high speed.

Following is a discussion as to how the benefits of high power density and high efficiency permanent magnet brushless dc motors can be realized in electric vehicle applications. Figure 7 shows a typical torque versus speed curve for a permanent magnet brushless dc motor. Note that the continuous torque line is relatively flat with stall torque being 40-50% greater than rated torque at high speed. Next, compare the available torque to the wheels from a permanent magnet motor with fixed gear reduction to the required torque based on an assumption that the motor and reduction design is such that the torque requirements are met at the half speed point. Such a curve would be as shown in Figure 8a. One can conclude that under the assumptions made, acceleration performance at lower speed would be marginal while at higher speeds more power would be available than needed for acceptable performance.

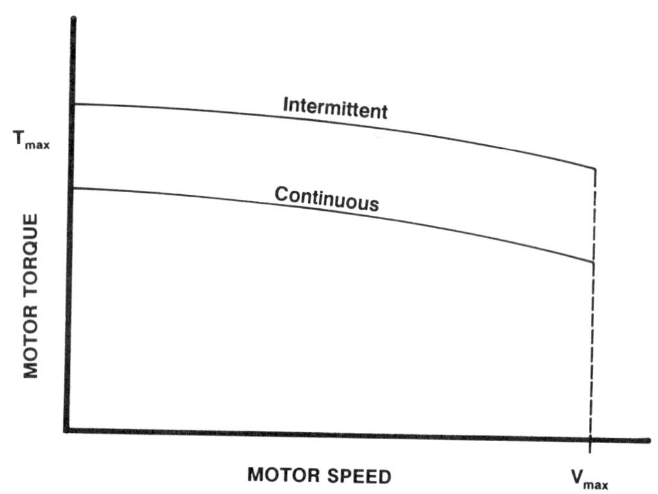

Figure 7

In a motor with a magnetic field that is constant, the available torque is directly proportional to the current delivered to the motor. Thus at low speed and for intermittent operation, it is possible for the permanent magnet motor of Figure 8a to meet vehicle requirements by increasing the size of the electronic controller and making more current available to the motor. The motor size may also be increased such that it is capable of satisfying torque requirements in continuous operation as opposed to intermittent operation. If one takes this approach as shown in Figure 8b, one now has a motor and electronic drive system that meets vehicle requirements but is considerably over sized for all other aspects of vehicle operation other than low speed acceleration. The cost of such a system is likely to be too high.

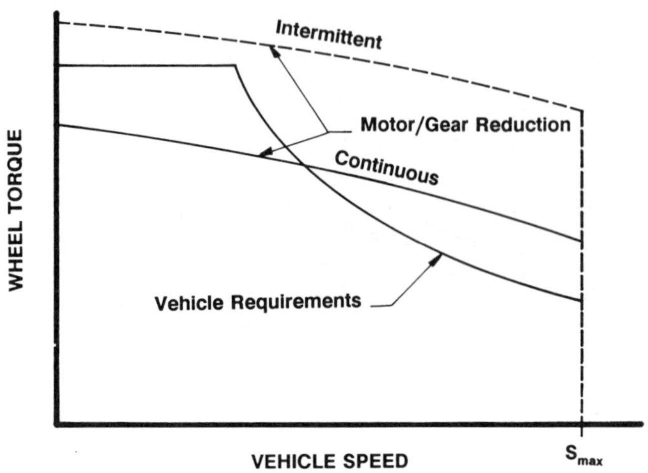

Figure 8a

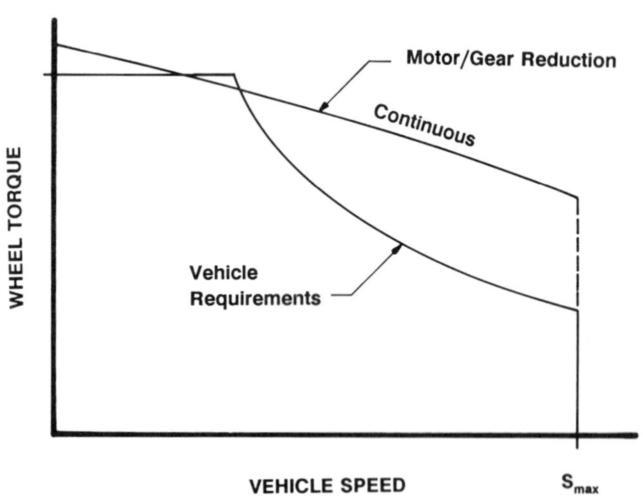

Figure 8b

There are several ways to get around this problem of low speed torque without increasing motor and electronic drive size, weight and cost. The most obvious solution is a multi-speed transmission. However, as suggested above, this is to be avoided if at all possible. Another approach is to use a continuously variable transmission or torque converter device. Technology in this area is lacking in terms of cost and efficiency. Unique currently is investigating two other techniques to increase the low speed torque capabilities of its permanent magnet motor for electric vehicle applications. The first approach involves use of field modification techniques. In this approach, the permanent field of the motor is altered by the use of an auxiliary field coil. At low speed, the total magnetic field can be increased to provide additional torque for constant current. For higher speed operation, the field is reduced and motor operates in a reduced torque overspeed condition.

Figure 9 shows the required wheel torque versus speed characteristic with a typical controller operating regime for a constant field permanent magnet motor and fixed ratio gear reduction. The upper limit of torque is determined by the maximum output current as shown by line I_{max}. The upper limit of speed is shown by V_{max} and is determined by the upper limit of system voltage. With these two curves we have now defined four regions of operation.

Region 1 defines an operating region that is within both the permanent magnet motor and controller continuous rating. The method of control in this region is pwm of the stator current. There is no field excitation in this region of operation. This is desirable for cruising below base speed as no power is wasted in the field coil. Region 2 is similar in operation but available power exceeds that actually required. Region 3 is the area where a combination of stator pwm and field strengthening is required to obtain desired torque. Region 4 is the overspeed region. In this region the stator pwm is reduced to zero and the field is weakened to increase and control speed.

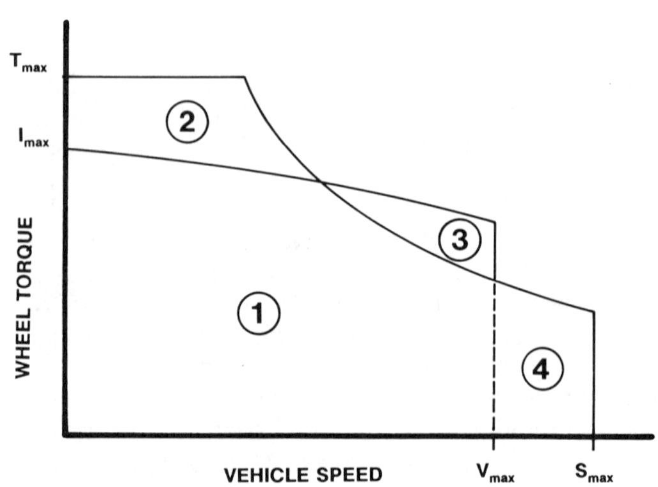

Figure 9

Practical implementation of the field modification concept has to accommodate changing conditions in the power system environment. Some sensitivities and their effects are:

Changing battery voltage - V_{max} varies

Controller temperature - I_{max} varies

Motor magnet temperature - I_{max} and V_{max} vary

Overall control is accomplished by sensing operational changes in the stator controller rather than absolute speed and torque values. This is desireable because torque is a very difficult parameter to obtain by present means of sensing. Speed must be used to a certain degree to determine the direction of field current applied. For example, at low speed, continued application of accelerator signal will cause the field to strengthen (with a corresponding decrease in stator pwm to maintain applied speed) in order to deliver higher torque. In the overspeed region, increasing accelerator pressure requires a decrease in the field, but not so rapidly that torque cannot be maintained. Using inexact means of implementing control, and considering the possibility that efficiencies might improve and transitions from one mode to another may be smoother if the boundaries of control overlap or are smeared, the resulting control map looks like Figure 10.

Another approach which holds promise for matching the brushless dc motor torque speed curve to the vehicle requirements involves a motor winding switch. Unique's motors are constructed with six slots per pole. In a 3-phase wye wound motor that are two parallel winding slots per phase leg. These parallel slots may be externally connected in either series or parallel. The concept is illustrated in Figure 11. In series operation, the torque constant of the motor is doubled and the top speed halved. In parallel operation the torque constant is reduced by a factor of two and the top speed doubles. The torque versus speed characteristics of a switchable motor are shown in Figure 12. Note that in this instance, the current handling capability of the electronic controller need not be increased to obtain higher lower speed torque. The switchable motor increases in size slightly as the stator thickness must be increased somewhat because of the higher winding resistance during series operation. In actual implementation, the winding switch gives the driver of a vehicle the sense of a very sudden drop off of torque. The feeling is similar to that one would get in shifting from second gear to fourth or fifth gear in a conventional vehicle. This torque variation may be accommodated by electronic smoothing of the torque-speed curve. It is necessary to embed a current/torque-speed algorithm in the electronic controller to smooth the transition and provide the desired vehicle torque versus speed characteristic.

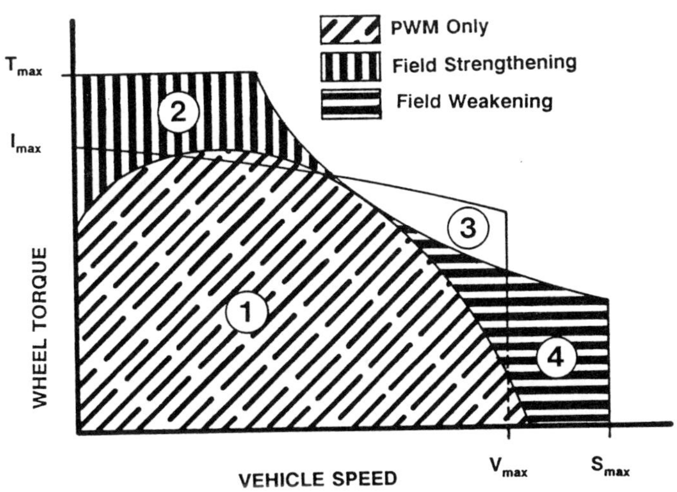

Figure 10

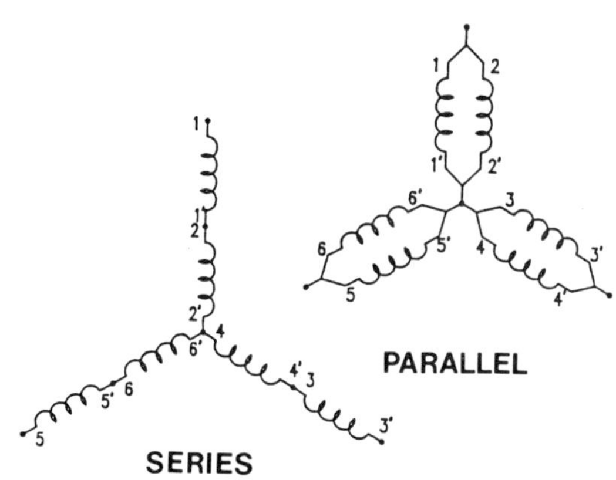

Figure 11

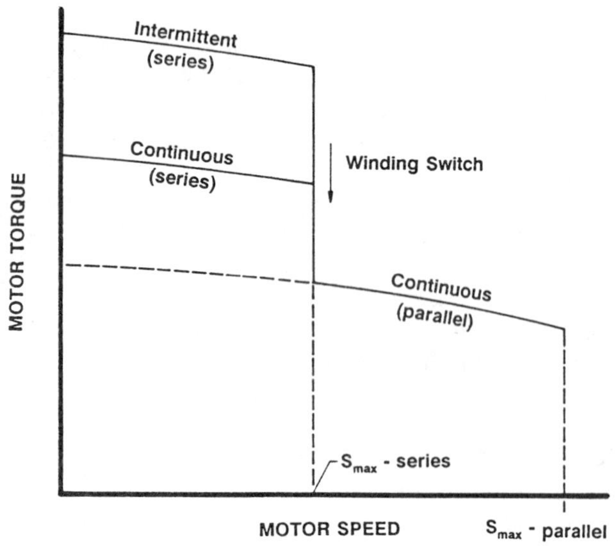

Figure 12

INTEGRATED MOTOR/TRANSAXLE

The compact size of the Unique motor permits its direct integration with a reduction gear and differential. Several approaches are possible. The Unique motor is hollow and depending on torque and gear ratio it may be possible to integrate the gear reduction directly into the body of the motor thus reducing overall size and system weight.

Typical gear ratios for electric vehicles will be in the 6 to 10:1 range depending on vehicle design and motor top speed. Except for relatively large diameter motors this may preclude packaging the gear reduction directly within the motor housing.

Figure 13 shows an integrated motor/transaxle which incorporates a 7500 rpm motor, a 7.5:1 gear reduction and a planetary differential which delivers the reducer output to both wheels of the vehicle by incorporating a hollow motor shaft.

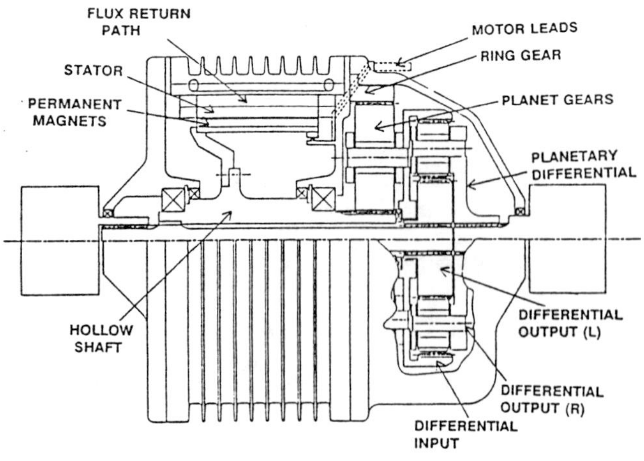

Figure 13

SUMMARY

Unique Mobility is optimizing it's advanced permanent magnet brushless dc motor technology for electric and hybrid electric vehicle application. The company is also exploring a wide variety of industrial applications for the technology.

The goal of achieving a cost effective brushless dc electric vehicle drive system which requires only a single stage of gear reduction requires adoption of new techniques to increase low speed torque. Means to achieve this currently being considered include field modification techniques and a two speed winding shift.

The motors can be integrated with gear reduction and differential to provide a lightweight, high efficiency completely integrated electric powertrain. Ultimately these technologies and other advances in the field will lead to an affordable electric transmission for vehicles. Unique's objectives in developing these concepts is to provide the basis for a commercially viable clean vehicle having greater range performance and economy than its fossil fueled counterparts.

910247

Electric Hybrid Drive Systems for Passenger Cars and Taxis

A. Kalberlah
Volkswagen AG
Wolfsburg, Germany

Abstract

Various hybrid drive configurations are described and their advantages and disadvantages for application in passenger cars are discussed; specifically, these are the series hybrid, the parallel hybrid, hybrid drives with added torque and speed, single and two-shaft hybrids.
The Volkswagen and AUDI group has developed different vehicles with hybrid drive for various applications. These vehicles are described and test results are presented on their energy consumption, emissions and driving performance. In conclusion, some considerations are pursued concerning their chances on the market in different scenarios.

1. INTRODUCTION

A hybrid drive, comprising both electric drive and internal combustion engine, can, if suitably designed, combine the advantages of the conventional vehicle drive system (large cruising range, good performance) with those of a purely electric drive (low noise and exhaust emissions, conservation of petroleum resources).

Vehicles equipped with such drive systems are thus far more flexible than electric vehicles; they are often just as versatile as vehicles with an internal combustion engine and consequently are not confined from the outset to the "second car" market.

Hybrid drives thus have far more extensive potential applications than electric drives; higher production rates could in principal therefore be achieved, leading to low manufacturing costs.

Surprisingly, there is a wealth of possibilities for realizing such hybrid drives. Some of these possibilities are outlined below and the advantages and disadvantages of these different designs are investigated by considering vehicles already manufactured.

2. SYSTEM ANALYSIS OF HYBRID DRIVES

2.1 SERIES HYBRID DRIVE - Taking the purely electric drive as a starting point, it is simple to conceive of a hybrid drive: the batteries of the electric vehicle are recharged when driving as required via generator driven by an internal combustion engine. This is not only simple in concept, but also in practice; several VW Electric Transporters, fitted with a suitable motor/generator assemply in the loading area, have been transferred from Wolfsburg to Essen. In actual fact, not only improvised vehicle drives have been built according to this principle, but also real drives, e.g. by Daimler Benz (1)* for a city bus.

The main advantage of the series hybrid:

It is possible to operate the internal combustion engine at a fixed operating point within its engine-speed/torque map. This point can be selected so that the engine functions with the greatest efficiency or produces particularly low emissions.

Nevertheless, the efficiency of the entire drive is not satisfactory. As the structure of this so-called series drive in Fig. 1 clearly shows, the 3 components V (internal combustion engine), G (generator) and E (electric motor) are arranged in series: the mechanical energy generated by the petrol

*Numbers in parentheses designate references at end of paper

engine is converted into electrical energy by the generator and this is again converted back into mechanical energy in the electric motor. Each process of conversion is afflicted with losses resulting in relatively bad efficiency. This is also confirmed by test results from a fleet of hybrid buses which were operated in Esslingen.

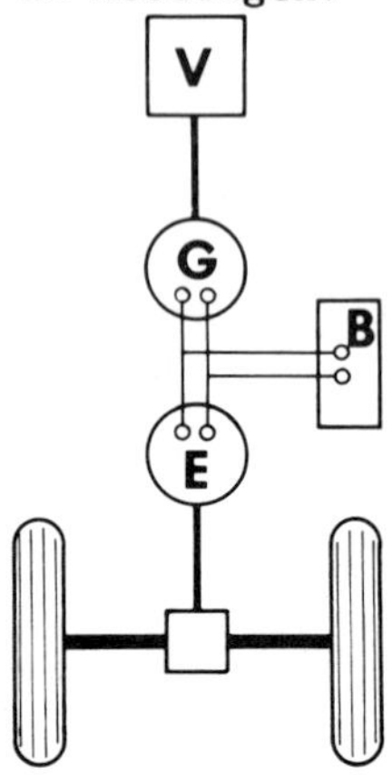

Fig 1 Structure of the series hybrid drive

A further disadvantage of this series hybrid drive is its heavy weight.

If power P_{max} is required at the drive axle e.g. for the maximum speed, the electric motor must be designed to produce this power P_{max}. If the driver wishes to drive long distances at this maximum speed, the power, which the battery could contribute, can be disregarded so that both, the generator and the internal combustion engine must be designed for the power P_{max} (because of the conversion losses in the electric motor and the generator, the power to be employed by the internal combustion engine would be even greater). A total power of over 3 P_{max} is required to drive at P_{max}. That makes this drive difficult to use in a universal vehicle (e.g. passenger cars) and expensive - particularly due to the two electric components.

For a vehicle, which only travels in the city, e.g. a delivery van or city bus, it may suffice, if the internal combustion engine and generator of the series hybrid are designed for average performance, since the battery could then provide for the power peaks. If this average performance is set at 1/2 P_{max}, it can be seen that, even in these urban vehicles, engines would have to be installed with power of more than 2 P_{max}.

Only if the combustion engine and the generator are very small compared to the electric motor - e.g. like in a range extender for electric vehicles (2), the series hybrid structure may be acceptable because additional weight, volumes and costs are small.

The problem of energy efficiency is neglectable in this case because most of the driving energy comes as electricity from the public network over the battery and the electric motor to the wheels and only a small amount comes out of the fuel tank over the IC-engine, the generator and the electric motor.

Another situation would be given, if the efficiency of the generator and the electric motor together are as good as the efficiency of the conventional gearbox. In this case, not only a series hybrid could make sense, but even a generator-motor-set instead of the mechanical gearbox. Some companies hope to be able to do this by using high speed synchronous generators and motors with new permanent magnets having a very high magnetic energy density (3).

But regarding Fig 2 showing that the efficiency of a mechanical gearbox is in a wide range over 90 %, it seems to be a long way to the "electrical gearbox".

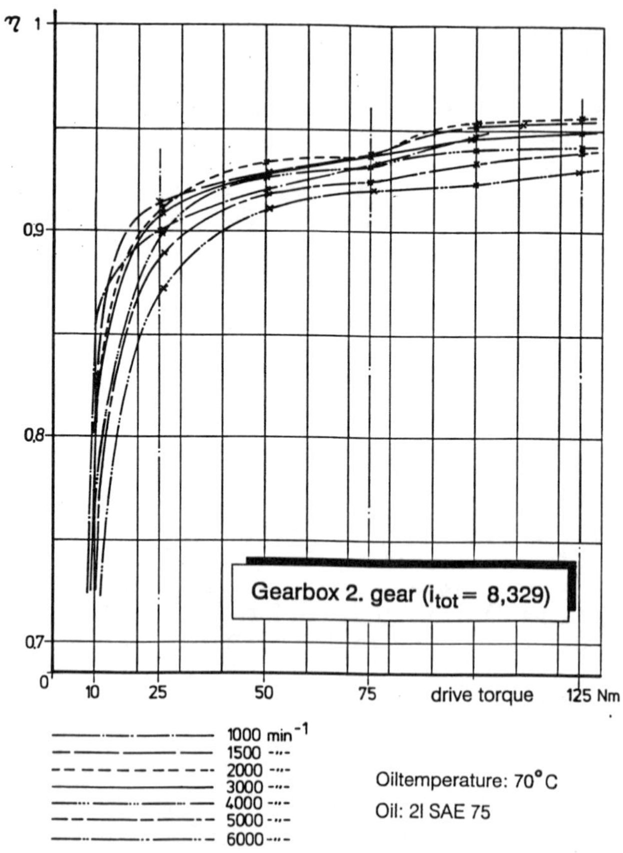

Fig. 2 Efficiency of a typical mechanical gearbox over the torque of the input shaft for different rotation numbers.

2.2 PARALLEL HYBRID DRIVES - Fig. 3 illustrates a parallel hybrid drive. Here, the internal combustion engine V and the electric motor E are not arranged in series (considering their power flow) but in parallel. The power provided by the two motors could be added to the vehicle drive. In this way, the power P_{max} required to drive a city vehicle, could be provided by designing e.g. both the internal combustion engine and the electric motor to give the power 1/2 P_{max} each. (For a series hybrid under the same conditions, at least 2 P_{max} would have to be installed.)

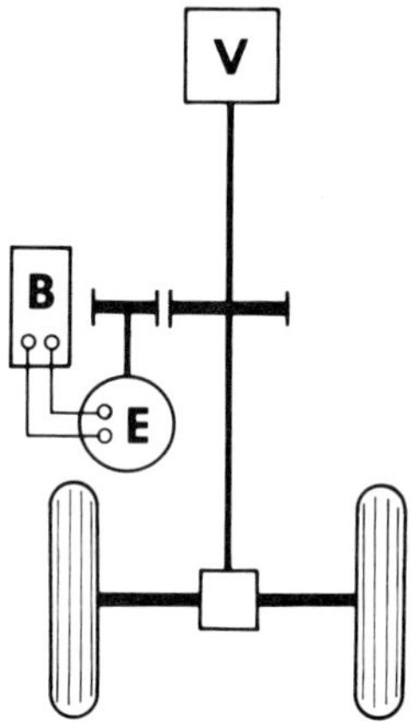

Fig. 3 Structure of a parallel hybrid drive

Whilst the series hybrid requires 2 electric machines, only one is present in the parallel hybrid. Nevertheless, even with the parallel hybrid, neither regenerative braking nor recharging the battery when driving should be disregarded, since the electric motor E can also function as a generator, e.g. if the total power of the internal combustion engine is not required to drive the axles.

Turning to a parallel hybrid for a universal vehicle requires the drive power P_{max} for longer periods, e.g. for long distances on the motorway. In this case, the power of the internal combustion engine must be P_{max}. The power of the electric motor could be selected from this completely independently. If P_{max} is selected for this too, since the series hybrid - if the battery power permits - can drive purely electrically at P_{max}, a total drive power of only 2 P_{max} needs to be installed. For a series hybrid under the same conditions, this would require more than 3 P_{max}.

In reality it is more sensible to select a much lower value for the power of the electric drive in such a parallel hybrid; i.e. a level which permits purely electric inner-city driving with acceptable performance. This is another point in favour of the parallel hybrid. The advantages of the parallel hybrid over the series hybrid can be summarized as follows:

- Improved efficiency and thereby lower fuel consumption in the internal combustion engine, since its mechanical energy is directly passed on to the drive axle. (Only if the battery is charged during driving - which should be avoided for reasons of conserving energy - is the same unfavourable efficiency chain present as in the series hybrid.)

- The generator is no longer required

- The weight is lower

- Costs are lower

2.3 STRUCTURES OF PARALLEL HYBRID DRIVES - Parallel hybrid drives can be realized in the most diverse forms, since the two mechanical power plants can be combined in various ways.

The version already discussed is illustrated at the far left of Fig. 4. Because of its two parallel-running drive shafts, we like to call this the two-shaft configuration.

In the next structure both drive units are arranged around a single shaft. As in the two-shaft configuration, the torques are added (or, in generator operation, subtracted), and this provides a free choice, within certain limits, in determining the extent to which the two units contribute to the total drive torque. This provides e.g. the opportunity to compensate for rapid changes in the desired torque with the electric motor torque and to permit only very slow changes in the throttle-valve setting in the internal combustion engine. This sluggishness in the throttle-valve has the effect of reducing exhaust emissions. However, the engine speeds are determined by the gearbox transmission. There is no opportunity for selection here.

The situation is reversed in the third structure illustrated in Fig. 4. Here, the power combination is performed by adding together the speeds of rotation of both drives in a differential gearbox positioned between the two units. There is, therefore, a certain freedom in dividing the speed between the two units, but the torques are fixed by the desired drive torques. In consequence, it is impossible to retard the throttle valve. A further disadvantage arises from the fact that the electric motor torque, on the one hand, and the petrol engine torque, on the other hand, must always be equal, although the torque-speed maps of both units are vastly different. Therefore, it is impossible to utilize e.g. the high torque of the electric motor at low speeds of revolution, which would basically be advantageous for traction.

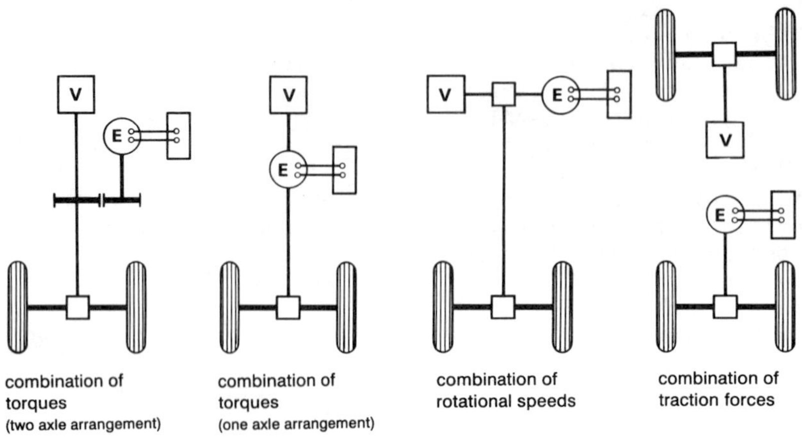

Fig. 4 Structures of parallel hybrid drives

For these reasons and because of the problems connected with the construction of a differential gearbox for combining high speeds with simultaneous high torques, we did not think it opportune to construct such hybrid drives with speed combination. Subsequently, the following chapter only presents hybrid drives with torque combination and with combination of traction forces.

This principle is illustrated by the structure on the far right in Fig. 4. It can be seen that there is no mechanical connection between the driving shaft of the combustion engine V operating on one axle and the electric motor E driving the other axle. This is a parallel hybrid too, because both motors could contribute in parallel to the traction force needed by the car.

3. CONSTRUCTED VW HYBRID DRIVES IN THE VW/AUDI-GROUP

Even if we limit ourselves to hybrid drives with torque combination, we can still get vastly differing drives depending on the type of vehicle and the application for which the drive is designed.

3.1 VW CITY TAXI WITH HYBRID DRIVE - The City taxi is a vehicle based on a VW bus (5,6). Its two-shaft hybrid drive is schematically represented in Fig. 5.

Fig. 6 shows an exploded view of the components. The power from the 37 kW IC-engine O is transmitted to the rear axle via a hydrodynamic converter W and a pneumatically operated clutch K. The electric motor torque is superimposed on the petrol engine torque via the fixed transmission.

The electric drive is relatively powerful with 16 kW continous output. With it, the taxi can be operated purely electrically in the innercities without presenting an obstacle to traffic. The petrol engine is started outside the city centre. The power plants of both the petrol engine and the electric motor are now available for acceleration. If the total power from the petrol engine is not required for driving, it is used to drive the electric motor which now functions as a generator. This serves to recharge the battery.

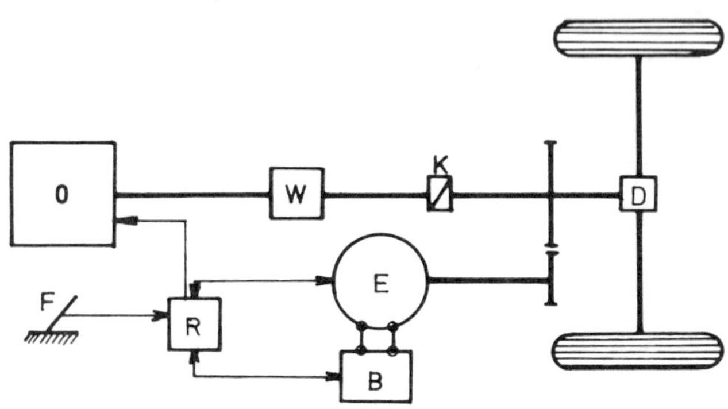

Fig. 5 Hybrid drive with torque combination in the VW City Taxi

Fig. 6 shows that this drive is able to function without the conventional transmission, which, of course, makes it very easy for the driver to use. He only has to select whether he wishes to drive in purely electric mode or hybrid, i.e. using both units, and express his wishes by operating the accelerator and brake pedals. A central control unit then controls the electric motor (via the electronic control) and the petrol engine (via a servo on the throttle valve), taking into account the battery charge.

In addition, the following functions are ensured:

- Protection of battery against complete discharge and overcharging
- Regenerative braking

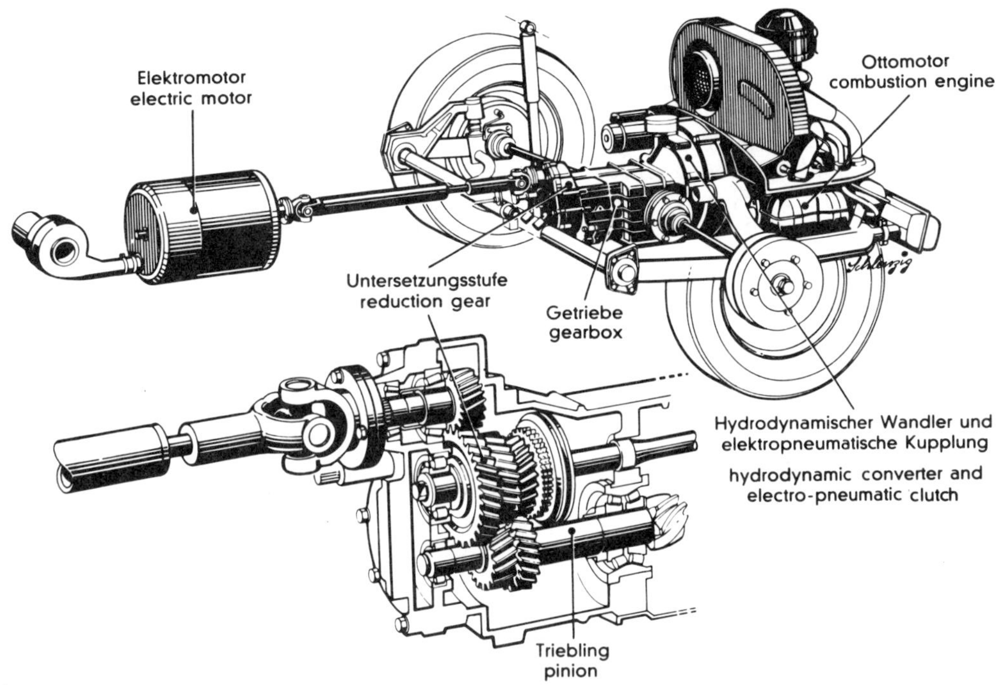

Fig. 6 Exploded view of the components of the VW City Taxi hybrid drive

- Sluggishness of the petrol engine in dynamic procedures by means of slow changes in the throttle valve setting. (The desired changes are temporarily taken over by the electric motor).

The vehicle has been tested over 100.000 km with the objective of obtaining information on petrol consumption and emissions. When the battery was mainly charged in driving (and not in the garage with a battery charger), petrol consumption rose by up to 25 % against a conventional VW Bus. The reason for this is the greater weight of the vehicle and the unfavourable efficiency chain for charging the battery in the vehicle. However, pollutant emissions dropped considerably in comparison to the conventional vehicle; specifically, CO by 27 % and HC by 85 %. If the operation of the electric motor as alternator is reduced (by limiting the charge current), fuel consumption and emissions fall (6).

The tractive resistance diagram in Fig. 7 shows that the vehicle displays respectable performance although all driving conditions are met in a single gear.

This is due firstly to the ideal, hyberbolic tractive force curve of the electric motor, secondly to the torque converter and thirdly to the principle of torque addition. This means that even a fully laden vehicle can increase its ability to climb by up to 23 %. In many areas, the tractive forces are even higher than if a conventional VW Bus is driven by a 37 kW petrol engine with 4-speed manual gearbox.

However, the limitations of this City Taxi drive should also be recognized; a mountain road with a gradient over 10 % of such length

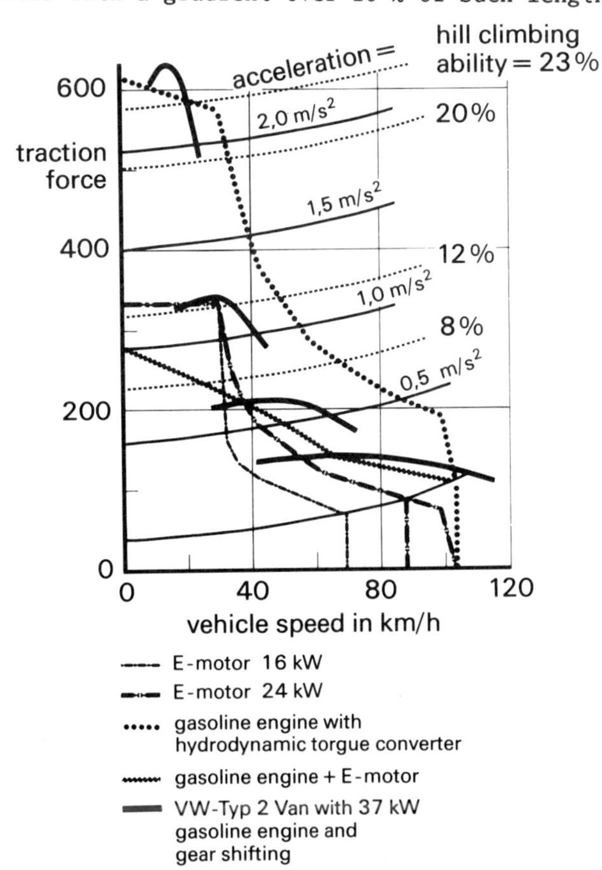

Fig. 7 Tractive resistance diagram of the VW City Taxi

that even an initially fully charged battery would not suffice, cannot be overcome with the petrol engine alone, since, in this type of operation, the max. gradient capability amounts to barely 10 %.

As much as this hybrid drive proved itself in the City taxi, it is, however, not suitable for use as a universal drive e.g. for passenger cars.

3.2 TWO SHAFT HYBRID DRIVE EVW2 - The objectives in developing this hybrid drive were:

a) Universal versatility
(Even with a flat battery the vehicle should be fully usable and still display approximately the same performance as a conventional vehicle.)

b) Economy with energy
(No battery charging in driving!)

c) Low additional costs in comparison to a conventional drive.

Since the requirements a and b point to an internal combustion engine of approximately normal power, the requirement c indicates that the additional electric drive should have the smallest power output possible; the question is however, what can you do with it. If e.g. a 5 kW motor is selected for a VW Golf, its power is sufficient to drive the vehicle on level ground at constant speeds up to 50 kph. This leads one to suppose that the petrol engine is often operated at very low power in inner city traffic. This is confirmed in Fig. 8: This represents the working pressure, a

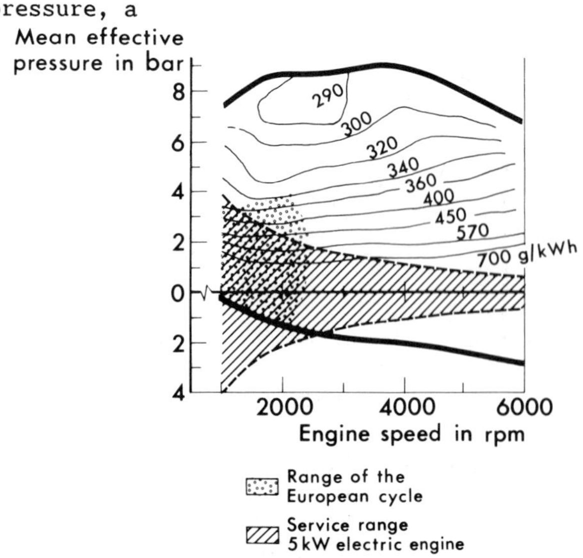

Fig. 8 Map of a petrol engine with: lines of constant consumption, working range of the petrol engine in the European Cycle, working range of a 5 kW electric motor

measurement of the petrol engine torque, as a function of engine speed. This map also contains the lines of constant specific consumption. The shaded area illustrates the range in which the petrol engine operates when the vehicle drives the European Cyle. It can be seen that very high specific consumptions, i.e. bad efficiency levels, are present here. It should be clear that the 5 kW electric motor which can cover the hatched area of the map, can take over a considerable proportion of these unfavourable operating conditions.

The petrol engine and electric motor complement one another very well in such a distribution of tasks, since each drive is allocated the tasks which it can perform particularly effectively: the petrol engine drive rapid acceleration and high speeds, the electric drive smaller loads, such as occur in innercity driving. In addition, lower energy consumption - the sum of petrol for the IC-engine and electrical energy for charging the battery - is to be expected. In 1983, the 1st prototype of this special hybrid design was manufactured in cooperation with the Electricité Neuchateloise S.A. and installed in a 1982 Golf (7).

The structure of the realized drive is schematically represented in Fig. 9; again, this is a two-shaft hybrid. The IC-engine is linked to the gearbox via an automatic electromechanical clutch K_O. The electric motor is connected to the other end of the gearbox inlet shaft via fixed transmission and an automatic, electromagnetic clutch K_E. In contrast to the City taxi, a conventional manual gearbox is present, this being a 4-speed semi-automatic. A central control unit regulates the use of the two drive motors, i.e. it decides which of the two motors should take over the current requirement for driving performance.

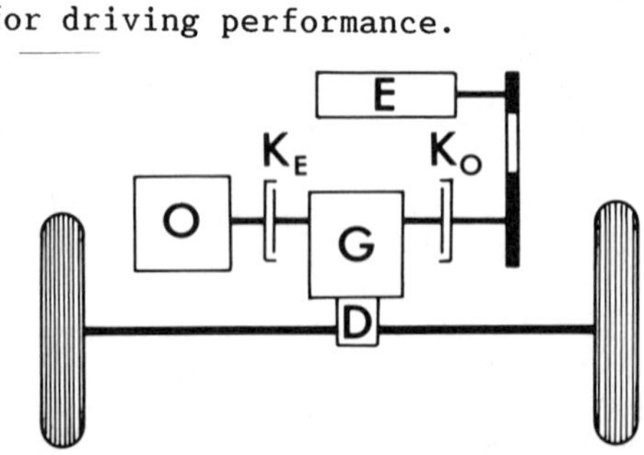

Fig. 9 Arrangement of the components in the EVW2 hybrid drive for a VW Golf

In operation of the realized drive is relatively simple. The familiar accelerator pedal alone is used to control and operate both motors and their clutches. There is no clutch pedal present. As in every conventional vehicle, the Hybrid Golf is started via the ignition lock. When the IC-engine has reached a sufficient operating temperature and the vehicle is standing still, both motors are switched off. When 1st gear is selected, the IC-engine is automatically started and the clutch between the IC-engine and gearbox is automatically closed when the accelerator pedal is depressed; the vehicle moves off and accelerates. Releasing the accelerator pedal (to zero) causes the IC-engine clutch to open and the next gear can be selected.

If the traffic situation requires less performance when driving with the IC-engine, the accelerator pedal should be lifted. If the accelerator pedal is fully released and no further action occurs for longer than 0.5 s. i.e. no renewed acceleration and no operation of the transmission selector lever, the IC-engine switches off and the vehicle now runs on without power. When the accelerator pedal is depressed again, the electric motor takes over the driving. If the electric motor's 5 kW drive power is not sufficient, because the driver wishes to go faster or accelerate, he should "give more gas", i.e. increase the accelerator-pedal travel beyond a certain point. The IC-engine is re-started so that the vehicle can drive on under the power of the IC-engine.

It is also possible to drive with either the electric motor alone or with the IC-engine alone:

If required, it would be possible to drive under purely electric power - albeit with reduced capacity for acceleration. At a constant speed of 50 kph, a distance of 36 km can be covered under purely electric power with the 200 kg lead-acid battery.

For longer journeys, e.g. holidays, the traction battery can be removed from the vehicle to obtain additional luggage space and increased loads. Driving is then done with the IC-engine.

During braking and downhill driving the vehicle's kinetic energy, which would otherwise be absorbed at the brakes or even result in additional petrol consumption on account of engine overrun, can be stored in the traction battery as electrical energy.

After completion of the vehicle, consumption and exhaust emission measurements for the European Cycle were first carried out on a roller dynamometer (8). For this, the percentage distribution of distance covered and time spent in each of the 4 operating conditions "Electric drive", "IC-engine drive", "Coasting" and "Stopping" were measured. These were intered in Fig. 10. It can be seen that the IC-engine is only used to cover 21.7 % of the distance, whilst the electric motor is used for 53.3 %.

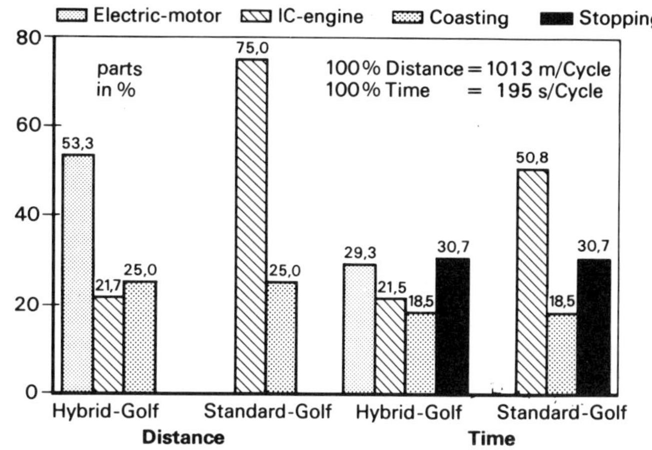

Fig. 10 Distance and time percentage distribution in the European Cycle with the EVW2 hybrid drive

This naturally has strong influence on reducing consumption and emissions. The Hybrid-Golf only consumes 33 % of the petrol which the conventional Golf required for the European Cycle. The quantities of CO and HC in the exhaust gases were greatly reduced, but the quantity of NO_x changed little.

Extensive road tests were then carried out in public traffic in Brunswick (7). The fuel saving in this case was not as high as in the European Cycle, but it still amounted to 40 % over the Standard Golf. The additional electrical energy required from the traction battery is 8.3 kWh/100 km or 13.8 kWh/100 km from the public supply network. At 0.20 DM/kWh that is around 2.80 DM/100 km.

In the Brunswick-Cycle, 60 km can be covered in hybrid driving with a single battery charge.

3.3 ONE SHAFT HYBRID DRIVE EVW1 - The aim of this project is to turn the basic concept of the EVW2 drive - optimal division of tasks between a relatively weak electric motor and a relatively powerful IC-engine - into reality in a considerably more compact construction suitable for motor cars.

The preliminary stage can be said to be the single-shaft hybrid drive jointly built by Volkswagen and General Electric and commissioned by the American Energy Ministy (DOE) (9).

The EVW1 hybrid drive for a VW Golf developed in the last few years by VW also incorporates a single-shaft configuration (10).

As Fig. 11 shows, both the IC-engine and the electric motor operate on the gearbox input shaft. The rotor of the electric motor replaces the gearbox flywheel and is flanked on each side by a clutch.

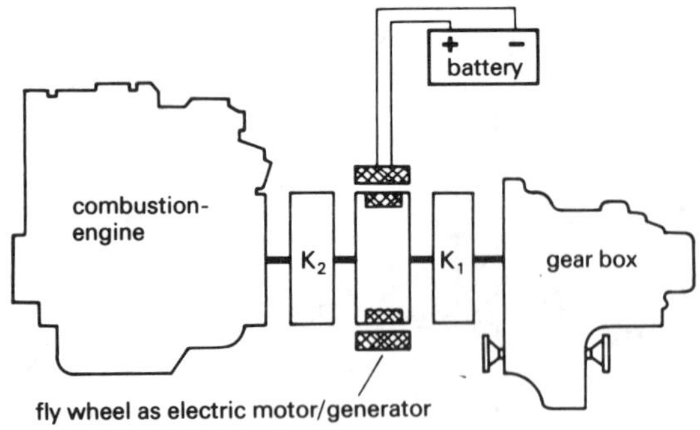

Fig. 11 Single-shaft hybrid drive EVW1 with disk electric motor between two clutches K1 and K2

To fit such a drive into vehicles at the lower end of the middle range, the electric motor must be constructed in disk form without commutators, as already indicated in Fig. 11. In cooperation with Bosch and LUK, VW has developed an asynchronous motor/generator in which the two clutches are integrated.

This compact construction means that the engine/gearbox drive unit only increases in length by 58 mm. The weight of the motor/generator, including the two clutches, in only 29 kg.

This ASM not only functions as the motor in conjunction with the 60 volt battery and as the generator, i.e. for regenerative braking, but it also replaces the conventional starter in the IC-engine; this is started by the ASM by closing K_2. For driving off, the 1st gear is selected; accelerating then closes the automatic clutch K_1. Switching through the further gears is done with clutch K_1 as in a semi-automatic. If the accelerator pedal is raised a long way, i.e. only a low power is required, K_2 opens, the IC-engine - now without flywheel stops immediately and the vehicle travels electrically with the asynchronous motor.

What is more, the electric drive also replaces the generator, since the 12 volt onboard circuit battery is charged when driving with the IC-engine. Apart from regenerative braking, the 60 volt battery is never charged via the "TANK -IC- engine - ASM as generator" chain, but always at the end of the journey with the aid of a battery charger from the 220 volt network. In this way, the most fuel can be saved.

A driving mode selector switch makes it possible to drive purely on the electric motor alone, i.e. without using the IC-engine at all, although the extremely low acceleration rate should be taken into account. This will be an important driving mode if we will have city regions where only cars with electric drive systems are allowed.

This drive concept has been put into practice once with a petrol engine and once with a diesel engine and a Golf has been equipped with each.

Both vehicles were tested in road traffic and on roller dynamometers.

Fig. 12 illustrates the emissions and the fuel consumption measured for the diesel/electric hybrid in the European City Cycle. For comparison, measured values from a standard diesel vehicle are also given. It can be seen that there is a strong reduction in emissions and a considerable saving of liquid fuels. For this only 16.3 kWh electrical energy per 100 km from the supply system must be used to charge the battery.

A fleet test with the EVW1 hybrid drive will be carried out in Switzerland starting in 1991. 20 VW-Golf with three different

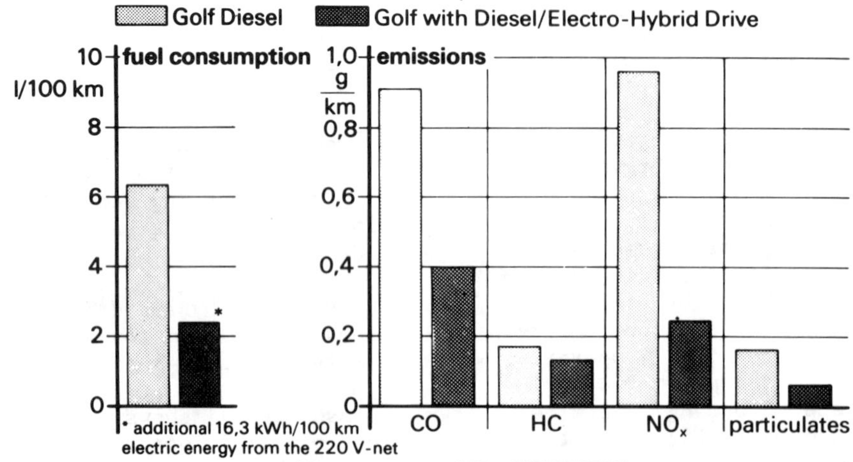

Fig. 12 Fuel consumption and emissions in the ECE Test Golf Diesel and Golf with diesel/electric hybrid drive

maintenancefree battery systems will be tested.

3.4 AUDI DUO HYBRID DRIVE - The principle of the AUDI DUO hybrid drive is illustrated by the structure on the far right in Fig. 4. This structure was chosen because the basic idea of the Audi duo is to interfere as little as possible with the standard layout of the normal front-wheel drive car by using the quattro rear axle to drive this electrically. To keep costs down the car is modified as little as possible.

Thus the Audi duo is a largely unmodified front-wheel drive vehicle. When required, for example in restricted inner-city zones, one can switch over to a 9.3 kW electric motor connected directly to the rear differential. So that hill starts are also possible the electric motor must have a high starting moment. This is provided by the series-wound motor, which gives the car a maximum speed of 50 to 60 km/h on level roads.

A comparison of the AUDI DUO and the EVW-Hybrid-Drive shows the following: If in future there should be individual traffic areas in which only electric drive is permitted, the Audi duo concept will have the better chances, because it is cheaper to produce in small series on account of the low level of vehicle modifications required. In a scenario where the saving of crude oil and the reduction of CO_2 emission are the main problems, the one shaft hybrid will be more likely to succeed. In large series production it would have the additional advantage that it is not dependent on having four driving wheels, and the induction motor replaces the flywheel and the starter motor of the combustion engine.

4. POTENTIAL FUEL SAVINGS WITH ELECTRIC VEHICLES AND ELECTRIC HYBRID VEHICLES

Besides the possibility of operating vehicles with electric and hybrid drive purely electrically, both types of drive offer the opportunity to save quantities of diesel and petrol by using electrical energy. If the electrical energy is produced from nonfossil primary energy there is given a possibility to reduce the CO_2 emission, too.

The following estimate should serve to determine the maximum potential saving: In 1981, a study initiated by the federal Ministry of Transport (BMV) of the usage and market potentials of battery-driven electric passenger cars resulted in a figure of 1.1 million vehicles in private households in the Federal Republic of Germany (11). The usage potential is that part of the total passenger cars which could be replaced by electric vehicles, considering the technical restrictions of the battery-electric drive. Further features, such as acceptance by the owner of the influence of vehicle costs on the decision to purchase, are of no significance for the usage potential. In 1981, this was approximately 5,6 % of the total number of passenger cars in households. Assuming that all vehicles travel an equal number of kilometers per year and consume an equal amount of petrol, a potential fuel saving to the same extent is achieved. In contrast to this, the hybrid drive is a universal drive. The usage potential is, therefore, identical to the total number of vehicles. If it can be assumed that each vehicle covers 32 % of its annual mileage in urban traffic (12) and the use of a hybrid drive saves 67 % of liquid fuels, this gives a potential fuel saving of 21,4 %. This figure is considerably larger than the potential saving of 5.6 % for electric vehicles, and, therefore, more attention should certainly be given to the hybrid drive than previously.

REFERENCES

(1) BADER, Ch. Vergleich alternativer Antriebssysteme für Omnibusse, Automobil-Industrie 4/82, pp. 461-465

(2) ANDERSON, W. Advanced electric vehicle powertrain components, Automative Technology Interntional 91, Sterling Publications International Ltd.

(3) HEIDELBERG, G. Magnet-Motor Antrieb für Nutzfahrzeuge und Verbrauchs- und Emissionsfreundliche Nahverkehrsbusse Elektrische Bahnen 87 (1989) p 135.

(4) MIERSCH, R. Otto-Elektro-Hybridantrieb im VW-City-Taxi, Electric Vehicle Council, Electric Vehicle Symposium 5, Philadelphia 1978 (872403)

(5) Electric- and Hybrid-Vehicle Powertrain Development, Fourth International Symposium on Automotive Propulsion Systems, April 17th - 22nd, 1977, Washington D.C.

(6) MIERSCH, R. 100.000 km testdrive of a VW-Microbus with an ICE electric-hybrid-drivesystem, EVS7, 26. - 29.06.1984, Versailles

(7) SARIDAKIS, N. Golf mit Otto-Elektro-Hybridantrieb, ATZ 87 (1985), pp. 581-584

(8) SARIDAKIS, N., JOSEFOWITZ, W. Petrol-Electric Hybrid Drive in the VW-Golf Electric Vehicle, Symposium - VIII, Washington D.C., 20. - 23.10.86

(9) BURKE, A.F., MIERSCH, R. Development of a full-size hybrid (electric/ICE) passenger car Electric Vehicle Development Group, 15. - 16.09.1981, London

(10) MIERSCH, R., SCHUSTEK, S., WIRTZ, R. VW Golf mit kompaktem Einwellen-Diesel-Elektro-Hybridantrieb, SIA Symposium, Paris 1987

(11) Einsatzbereiche sowie Anwendungs- und Marktpotential von batteriebetriebenen Elektro-Pkw im Straßenverkehr, Forschung Stadtverkehr, Heft 32, 1983, Published by BMV, Hoermann Verlag, Hof

(12) Verkehr in Zahlen 1983, Bundesminister für Verkehr

910248

Natural Gas Hybrid Electric Bus

Alan T. Gilbert
Unique Mobility Inc.
Englewood, CO

Richard Gunn
Ontario Bus Industries
Mississauga, Ontario

ABSTRACT

The design and predicted performance of a hybrid electric powered transit bus is described. The bus is a 7.6 meter (25 ft), 24 passenger vehicle that incorporates a low floor design and rear door accessible to handicap passengers. The low floor and rear door are made possible by the use of individual high power density permanent magnet motors driving the rear wheels. The hybrid electric drive system consists of a compressed natural gas fueled internal combustion engine that drives a generator which in conjunction with storage batteries supply power to the two traction motors.

INTRODUCTION

In response to the increasingly stringent emissions requirements and the escalating fuel costs a hybrid electric powered bus is being developed and built.

The objective of the hybrid electric bus project is to prove a concept. The concept is that an I.C. engine/generator, providing steady state power, and augmented by storage batteries to supply the power required above the engine/generator base level, will provide lower emissions and increased fuel economy as compared to a conventional powered diesel bus. The batteries are to be rechargeable via recouping a portion of the braking energy and by the engine/generator during periods of low driving / auxiliary energy requirements.

Through an energy management system the engine/generator and storage batteries will be connected to two traction motors at the rear wheels.

The project is divided into 4 stages:

1. Develop a computer modelling program to predict performance and to assist in selection of the drivetrain components.

2. Research engine emissions with the engine operating at a steady state.

3. Detailed design and construction.

4. Vehicle testing and analysis.

In order to predict the performance and evaluate the options of major components a computer model was developed. The vehicle and component parameters are entered into the computer model as input variables, and the model automatically calculates the vehicle performance, range and the performance of the individual components.

COMPONENT SELECTION AND PERFORMANCE PREDICTIONS

SELECTION OF THE BUS CHASSIS- The bus chassis chosen is an Orion II 7.6 meter (25 ft). The bus is equipped with a continuous low floor design of 305 mm (12.0 inches) from the ground and rear wheel independent suspension. The air conditioning system and the compressed natural gas fuel tanks are located in the roof of the bus. The engine/generator is to be located in the front of the bus. The batteries are to be located over the rear wheel wells and inside the engine compartment. The bus is also capable of carrying the added electrical equipment and passengers. The gross vehicle weight rating is 10,900 kilograms (24,000 lbs). A plan and elevation view of the bus indicating the location of the drivetrain components are shown in figure 1. The drivetrain components are located to give equal weight distribution between front and rear axles.

INTERNAL COMBUSTION ENGINE- The IC Engine for the bus is to be a 4.3L V6 rated at 67.0 kW (90.0 hp) at 3000 rpm converted to operate on compressed natural gas. The engine will drive all the accessories through belts at the front of the engine and drive the generator through a speed increaser planetary gear drive at the flywheel.

The engine is to be controlled by an Engine Management System (EMS) that allows the engine speed and power output to vary with load conditions. However the problems associated with transient speed changes, poor fuel economy and high emissions, will be minimized by the fact that the EMS controls the rate at which the speed increases. The EMS will monitor the load conditions and signal the engine to increase the speed and power output at a slow rate to allow the fuel and ignition systems to maintain a stabilized operation; therefore, increasing fuel economy and decreasing emissions.

As mentioned above the engine is a converted automotive engine; therefore, an emission control system has been developed and is available at a reasonable cost. This emission control system will be used in conjunction with the Engine Management System to further decrease the overall emissions of the bus.

Based on the engine power output and control system the bus will have the ability to operate on the typical downtown stop and go cycle with no decrease in net battery energy and possibly an increase in the stored energy because of charging during low power requirement periods an regenerative braking.

DRIVETRAIN CONFIGURATION- The bus is to be rear wheel drive with the tractive power supplied by two (2) 70.0 kW motors combined with a single planetary gear reduction of 2.77:1 ratio that is directly coupled to the drive wheels through a second planetary gear reduction of 5.20:1 included in the wheel hub. The overall reduction in the two (2) reduction units equals 14.40:1 giving a maximum obtainable speed of 88.5 kilometers per hour (55.0 mph). The rear suspension is to be an independent trailing arm design with the traction motors mounted directly to the trailing arms. This design allows the low floor level to be maintained between the rear wheel wells. The bus is currently designed with an independent rear suspension so the mounting points are already in place therefore simplifying the design and installation.

The differential effect for the independent rear wheel drive of the bus is to be controlled by the motor controllers. The controllers will monitor the steering wheel position and wheel speed in order to determine the correct values of wheel speed and torque.

The power is distributed differently depending in which mode the vehicle is operating. The modes of operation are shown as power flow charts in figures 2 through 4.

OPERATION ON IC ENGINE POWER (figure 2)- Steady state operation on power produced by the IC engine only. In this mode the engine supplies all the power required to propel the vehicle and drive the accessories as seen in figure 6. The present selection of components allows the bus to operate at ~ 62 kilometers per hour (37 mph) without requiring power from the storage batteries as seen in figure 5.

OPERATION ON IC ENGINE AND BATTERY POWER (figure 3)- This occurs during steady state operation above 62 kmph, acceleration, and climbing a grade. The IC engine is operating at full power output to supply the accessories and a portion of the power to propel the vehicle. The remainder of the power to propel the bus must be provided by the storage batteries. The amount of power that the storage batteries must supply for the varying conditions is shown in figures 5, 6, and 7.

REGENERATIVE BRAKING WITH IC ENGINE OPERATING (figure 4) - During this mode the IC engine provides power to the accessories only, and power generated by the regenerative braking is delivered to the battery for storage to be used later.

STORAGE BATTERIES- The storage batteries are primarily to supply power during acceleration, on a grade, and at high speed cruising as mentioned previously. The batteries to be used are to be nominally 12 volts with a capacity of 160 amp-hours and a weight of 64.0 kg (141.0 lb). Fifteen (15) batteries are to be connected in a series string. Two (2) strings are to be connected in parallel to yield a system of 180 volts and 320 amp-hr capacity.

The battery pack will also be used to store energy that is recovered during regenerative braking. This arrangement is beneficial in two (2) ways, first it extends the range by adding to the usable on-board energy, secondly it increases the brake lining life.

PREDICTED RANGE- The computer model developed, titled Electric Vehicle Computer Model, was used to predict the usable range of the bus operating on different representative duty cycles. The predictions were made with the bus loaded to the gross vehicle weight rating of 10900 kilograms (24000 lb). The duty cycles selected to represent the typical operation of the Hybrid Electric Bus are listed below with a brief description;

1. 80/0, 80.0 kilometers per hour (50.0 mph) on level ground.

2. 70/2.5, 70.0 kilometers per hour (44.0 mph) on a 2.5% grade.

3. 12/12.5, 12.0 kilometers per hour (7.5 mph) on a 12.5% grade.

4. CBD, Central Business District cycle taken from the UMPTA White Book shown in figure 8.

5. Arterial, a simulation of typical arterial driving taken from the UMPTA White Book as seen in figure 8.

TABLE 1 PREDICTED RANGE.

CYCLE	RANGE (km)	LIMITING FACTOR
80/0	128.6	BATTERY
70/2.5	32.3	BATTERY
12/12.5	14.3	BATTERY
CBD	318.6	FUEL
ARTERIAL	104.3	BATTERY

The range of the hybrid operation is based on 131.5 kilograms (290 lb) of CNG fuel on board and the batteries at a beginning state of charge of 80.0% and at an ending state of charge of 20.0%. The Electric Vehicle Model calculates the range based on the fuel and on the battery capacity the lesser of the two is reported as the maximum achievable range.

The energy consumed per kilometer for the hybrid electric and diesel bus is compared in table 2 for the respective duty cycles.

TABLE 2

On-Board Fuel Energy Consumption per Kilometer (kW-hr/km)

CYCLE	HYBRID (kW-hr/km)	DIESEL (kW-hr/km)	SAVINGS %
80/0	2.5	3.4	26
70/2.5	2.9	5.9	51
12/12.5	16.9	19.1	11
CBD	5.7	6.1	6
ARTERIAL	4.0	5.1	22

*Based on an IC engine Brake Specific Fuel Consumption of 2.25 kW/Bhp (7720 BTU/Bhp-hr).

From the information given in table 2 it can be seen that a hybrid electric bus will have a decreased energy consumption per kilometer in all representative duty cycles.

The fact that the Electric Vehicle Model predicted that the hybrid electric bus has a decreased energy consumption was a major positive step in proving one of the primary objectives of the project.

POWER REQUIREMENTS- Based on the IC engine maximum power output and the accessories operating at 100 percent the hybrid design allows the IC engine to supply all the power to propel the bus and operate the accessories at speeds less than 62.0 km/hr at speeds over this break even value the storage batteries supply the make up power as shown in figure 5.

For a full throttle acceleration the IC engine will supply all the vehicle power required up to 18 km/hr as seen in figure 6. During the acceleration the IC engine supplies 45% of the total energy expended for the full throttle acceleration from 0 to 88.5 km/hr. The IC engine will supply all the power required to drive and power the accessories at a constant speed of 12 km/hr on grades equal to and less than 7.0% as seen in figure 7. The battery power required for a 12.5% grade and 12 km/hr is equal to 86%.

ENGINE EMISSIONS

With the limited research that has been done to date, there is an indication that by running the engine at a steady state condition and allowing the batteries to supply the energy for the peak loads the engine emissions may be reduced by a significant amount, as compared to a conventional diesel engine. More definitive results will be obtained after the emissions testing that is scheduled for the end of the first quarter of 1991.

DETAILED DESIGN AND CONSTRUCTION

The particulars on the detailed design are not available at this time as the project is currently involved in this stage. Most items are yet to be finalized. Since the detailed design stage is not completed the construction of the bus is not started. The construction of the bus will involve the incorporation of the electric drive system into an existing bus chassis.

VEHICLE TESTING AND ANALYSIS

Upon completion the bus will be subjected to a series of tests and evaluations to refine and optimize the control system, including Engine Management and the brake regenerative systems. The fossil fuel consumption will be measured and compared to that of a diesel powered bus to determine if the hybrid electric drive system is actually beneficial, by increasing fuel economy and decreasing the emissions of the vehicle.

CONCLUSIONS

Based on the computer model predictions, the concept that the combination of the IC engine/generator providing steady state power and the storage batteries to supply power during peak demand periods and the ability to recover and store energy during braking, the bus will generate lower emissions and increased fuel economy than a comparable diesel fueled bus. The knowledge gained on the concept vehicle will be applied to larger buses in the 9.8 to 14.5 meter (30 to 45 ft) long range. The 9.8 meter buses will incorporate a four wheel drive system and the 13.1 meter and longer buses will have a six wheel drive system with two drive axles at the rear of the vehicle. The final design of the hybrid electric buses is to have all accessories electrically driven so that the bus can operate on battery power only.

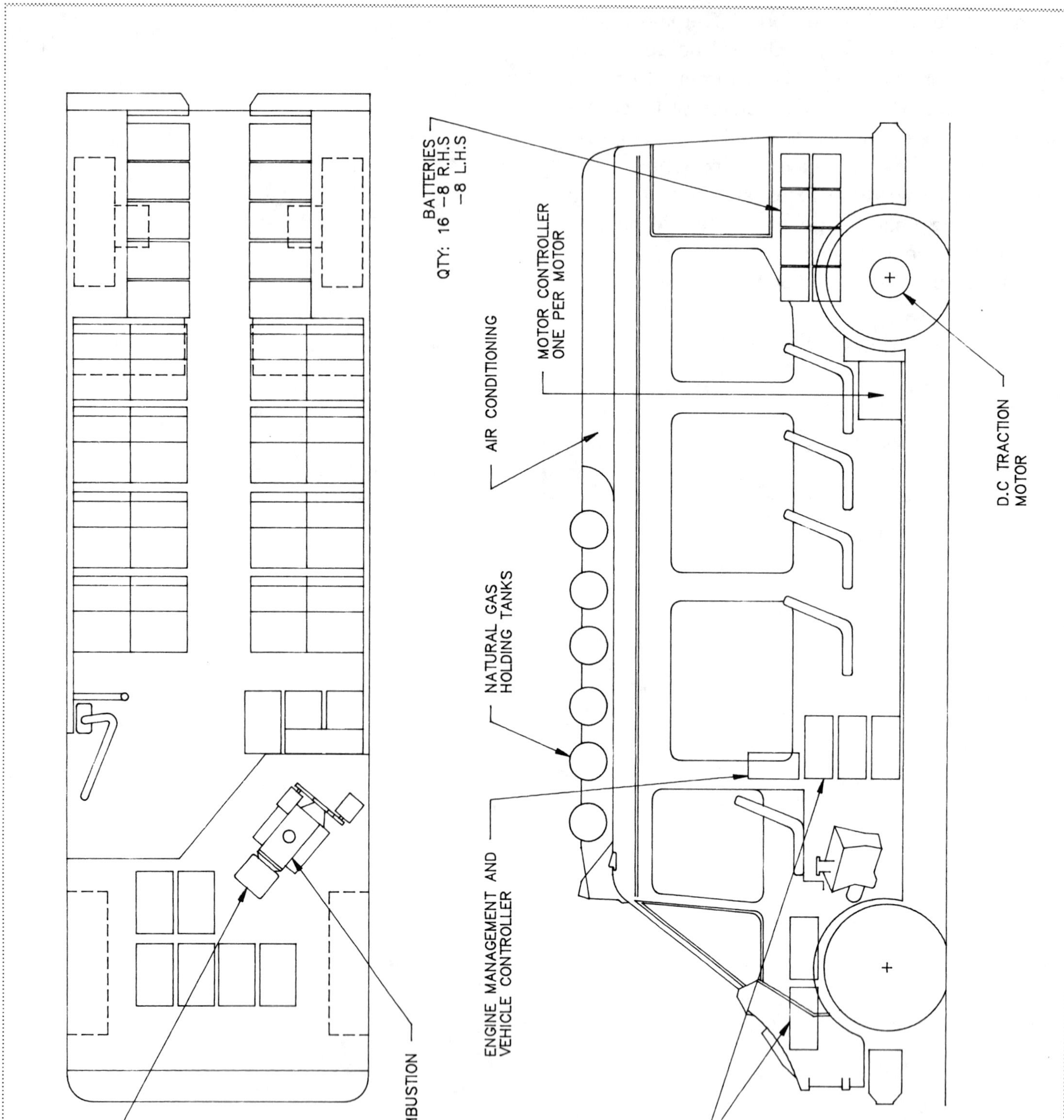

GENERATOR
INTERNAL COMBUSTION ENGINE
BATTERIES QTY: 14
ENGINE MANAGEMENT AND VEHICLE CONTROLLER
NATURAL GAS HOLDING TANKS
AIR CONDITIONING
MOTOR CONTROLLER ONE PER MOTOR
BATTERIES QTY: 16 –8 R.H.S –8 L.H.S
D.C TRACTION MOTOR

Figure 1

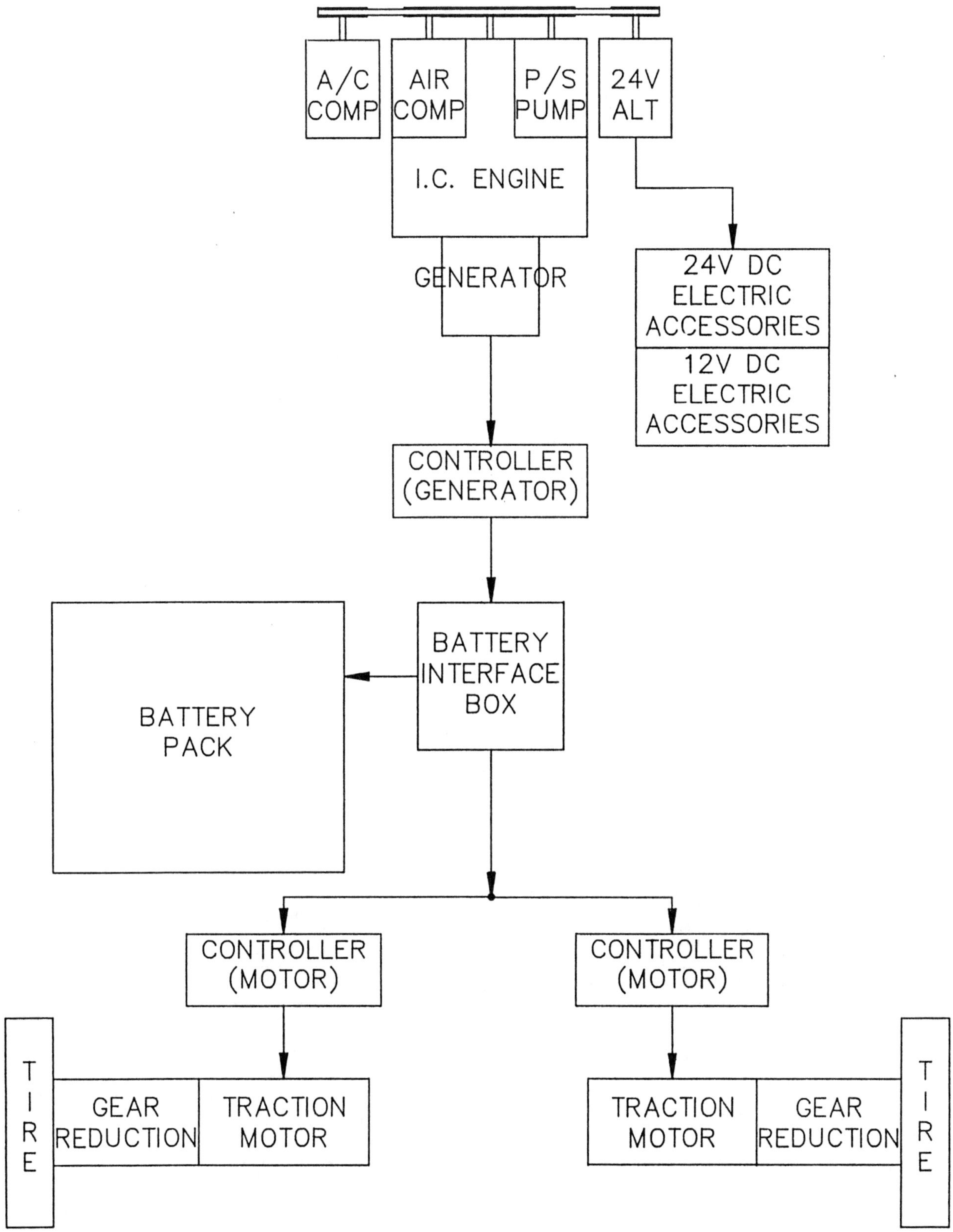
A/C COMP
AIR COMP
P/S PUMP
24V ALT
I.C. ENGINE
GENERATOR
24V DC ELECTRIC ACCESSORIES
12V DC ELECTRIC ACCESSORIES
CONTROLLER (GENERATOR)
BATTERY PACK
BATTERY INTERFACE BOX
CONTROLLER (MOTOR)
CONTROLLER (MOTOR)
TIRE
GEAR REDUCTION
TRACTION MOTOR
TRACTION MOTOR
GEAR REDUCTION
TIRE

Figure 2

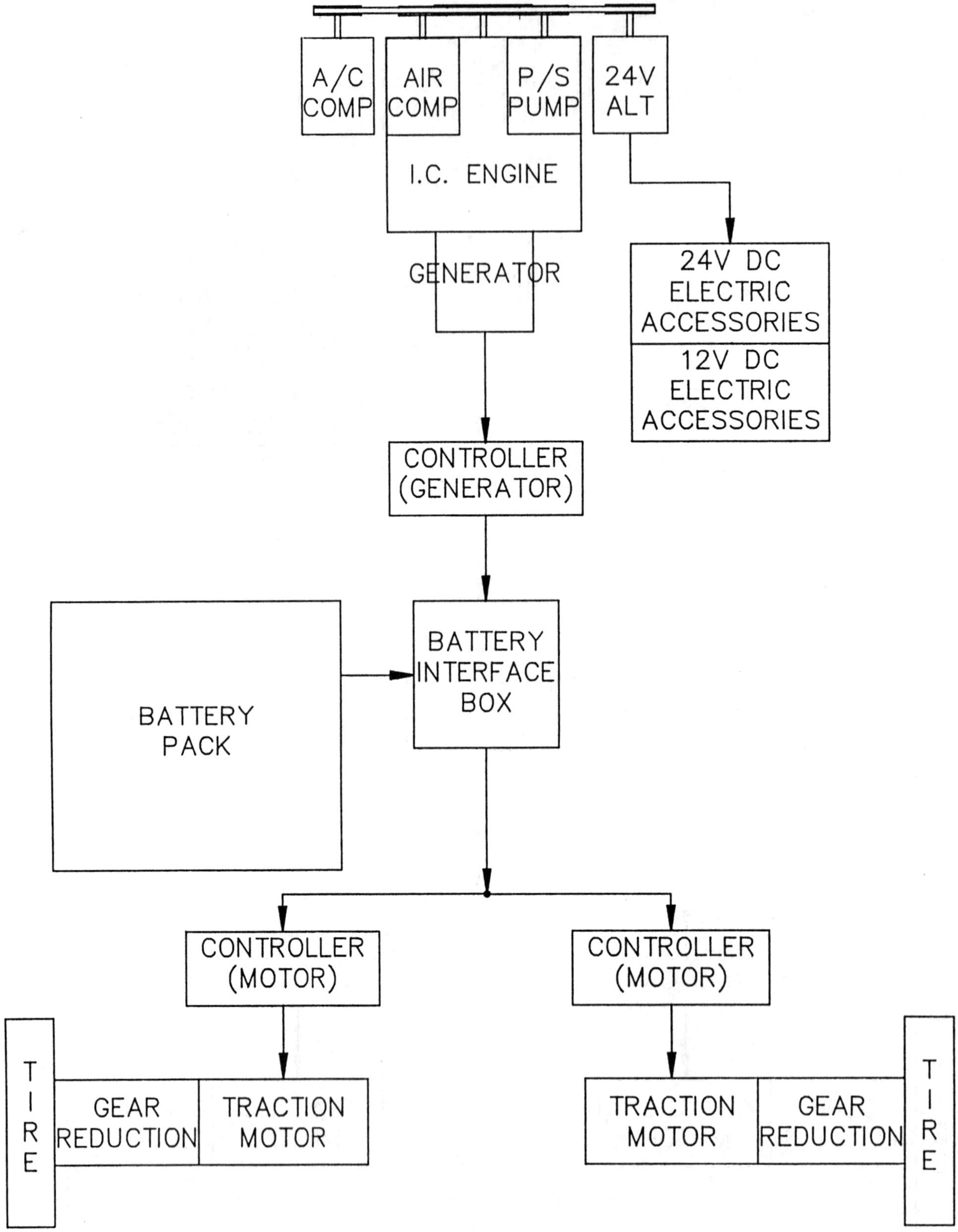

A/C COMP
AIR COMP
P/S PUMP
24V ALT
I.C. ENGINE
GENERATOR
24V DC ELECTRIC ACCESSORIES
12V DC ELECTRIC ACCESSORIES
CONTROLLER (GENERATOR)
BATTERY PACK
BATTERY INTERFACE BOX
CONTROLLER (MOTOR)
CONTROLLER (MOTOR)
TIRE
GEAR REDUCTION
TRACTION MOTOR
TRACTION MOTOR
GEAR REDUCTION
TIRE

Figure 3

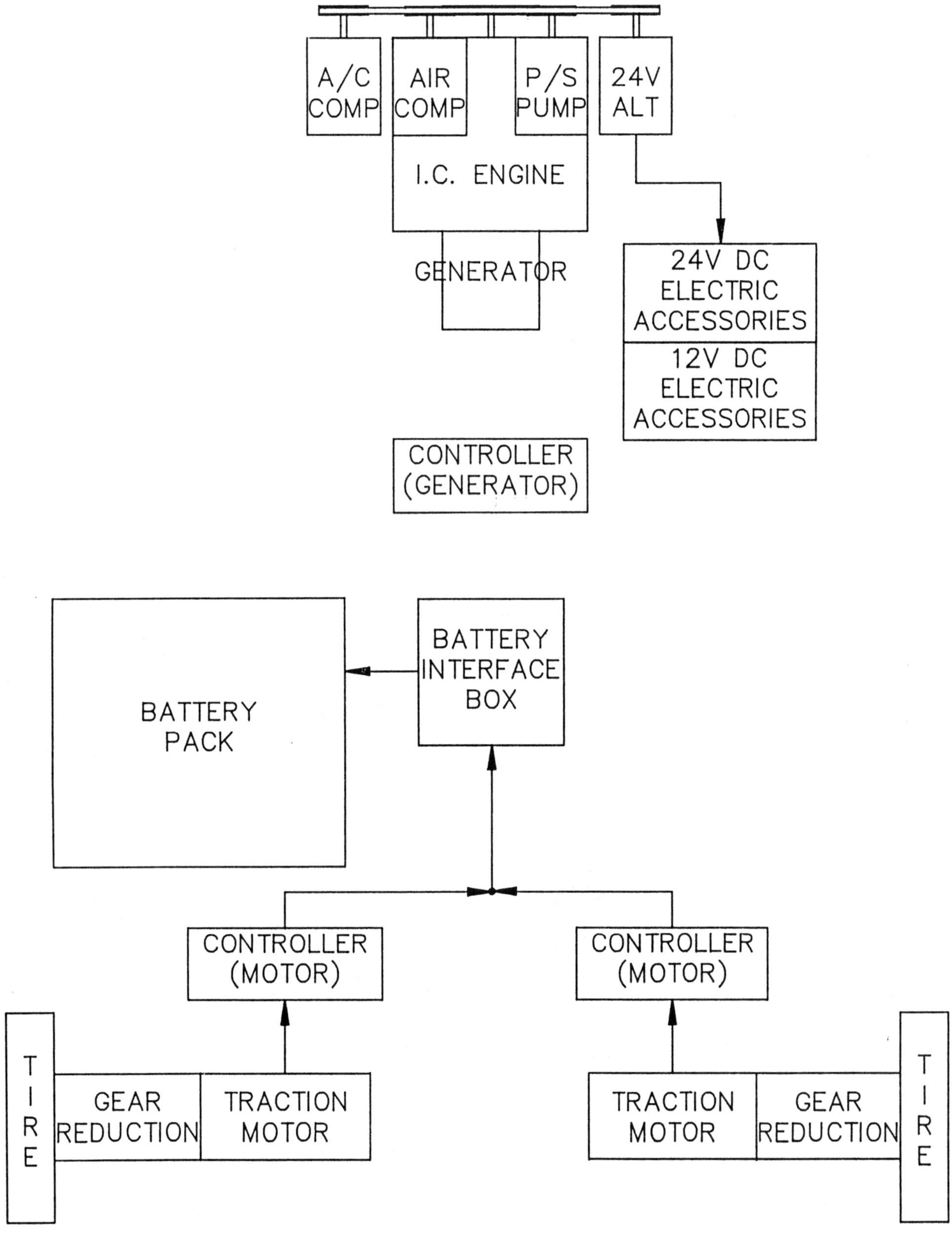
A/C
COMP
AIR
COMP
P/S
PUMP
24V
ALT
I.C. ENGINE
GENERATOR
24V DC
ELECTRIC
ACCESSORIES
12V DC
ELECTRIC
ACCESSORIES
CONTROLLER
(GENERATOR)
BATTERY
PACK
BATTERY
INTERFACE
BOX
CONTROLLER
(MOTOR)
CONTROLLER
(MOTOR)
TIRE
GEAR
REDUCTION
TRACTION
MOTOR
TRACTION
MOTOR
GEAR
REDUCTION
TIRE

Figure 4

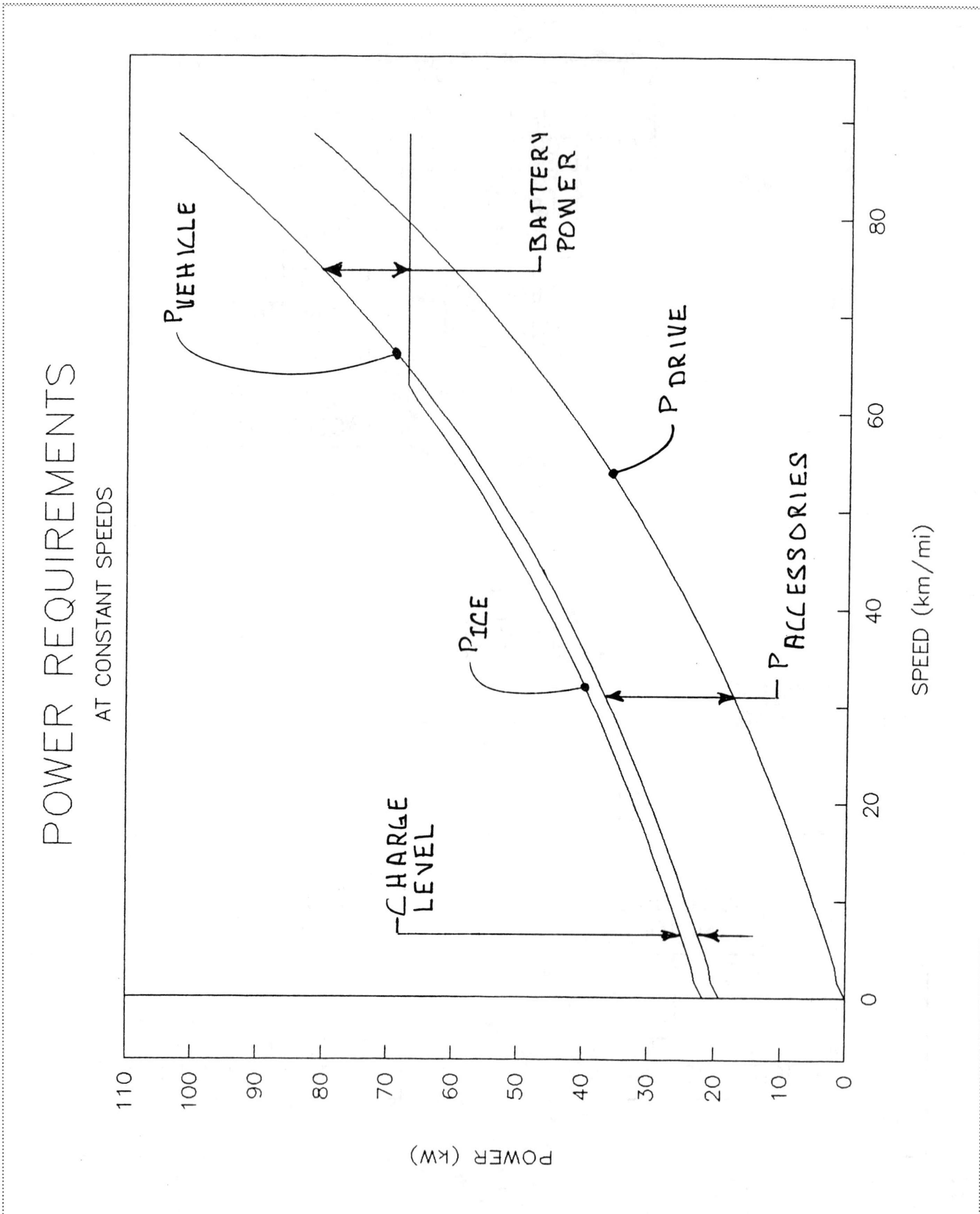
POWER REQUIREMENTS
AT CONSTANT SPEEDS
POWER (kW)
SPEED (km/mi)
P VEHICLE
BATTERY POWER
P DRIVE
P ACCESSORIES
P ICE
CHARGE LEVEL
0
10
20
30
40
50
60
70
80
90
100
110
0
20
40
60
80

Figure 5

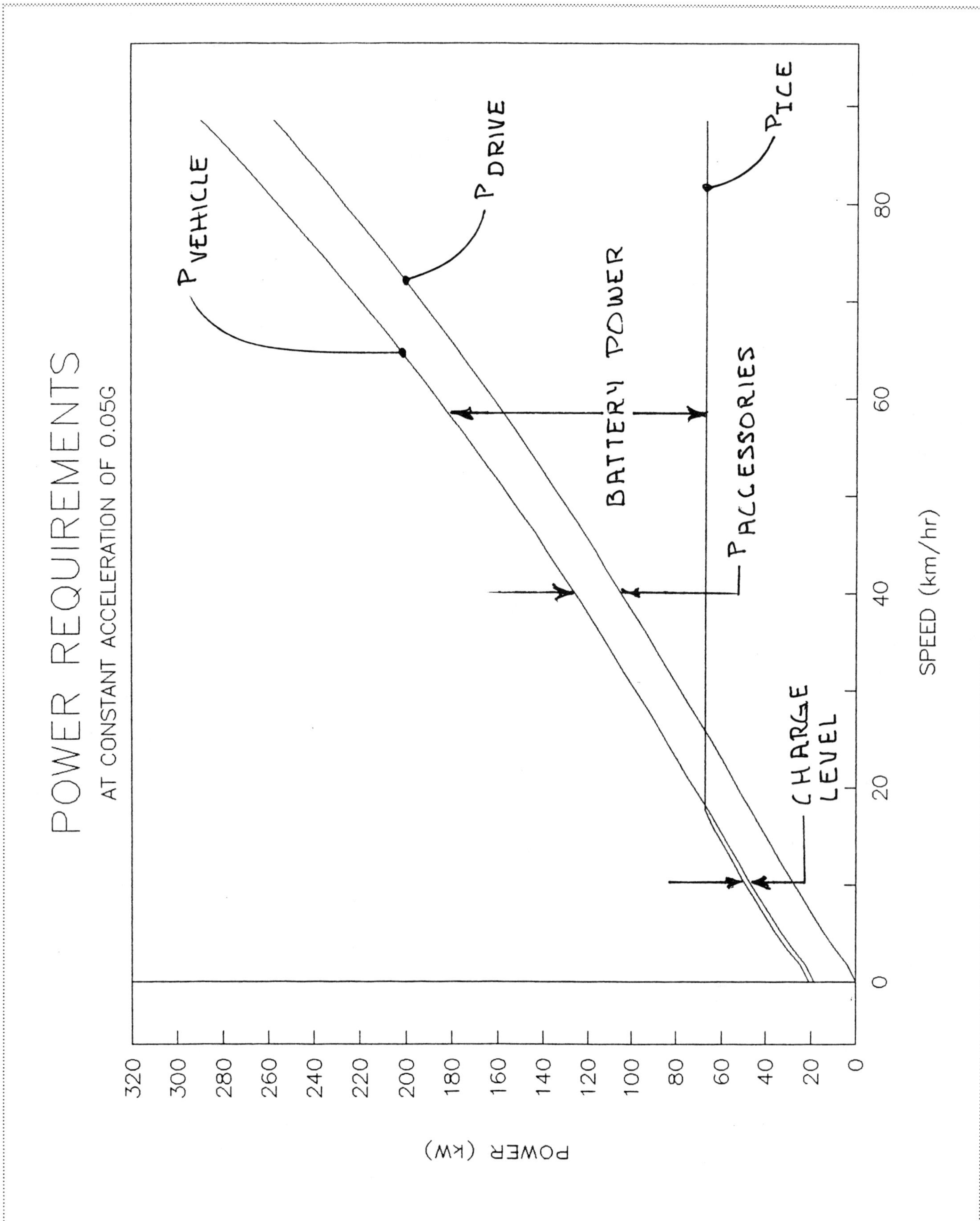
POWER REQUIREMENTS
AT CONSTANT ACCELERATION OF 0.05G
$P_{VEHICLE}$
P_{DRIVE}
P_{ICE}
BATTERY POWER
$P_{ACCESSORIES}$
CHARGE LEVEL
SPEED (km/hr)
0
20
40
60
80
POWER (KW)
0
20
40
60
80
100
120
140
160
180
200
220
240
260
280
300
320

Figure 6

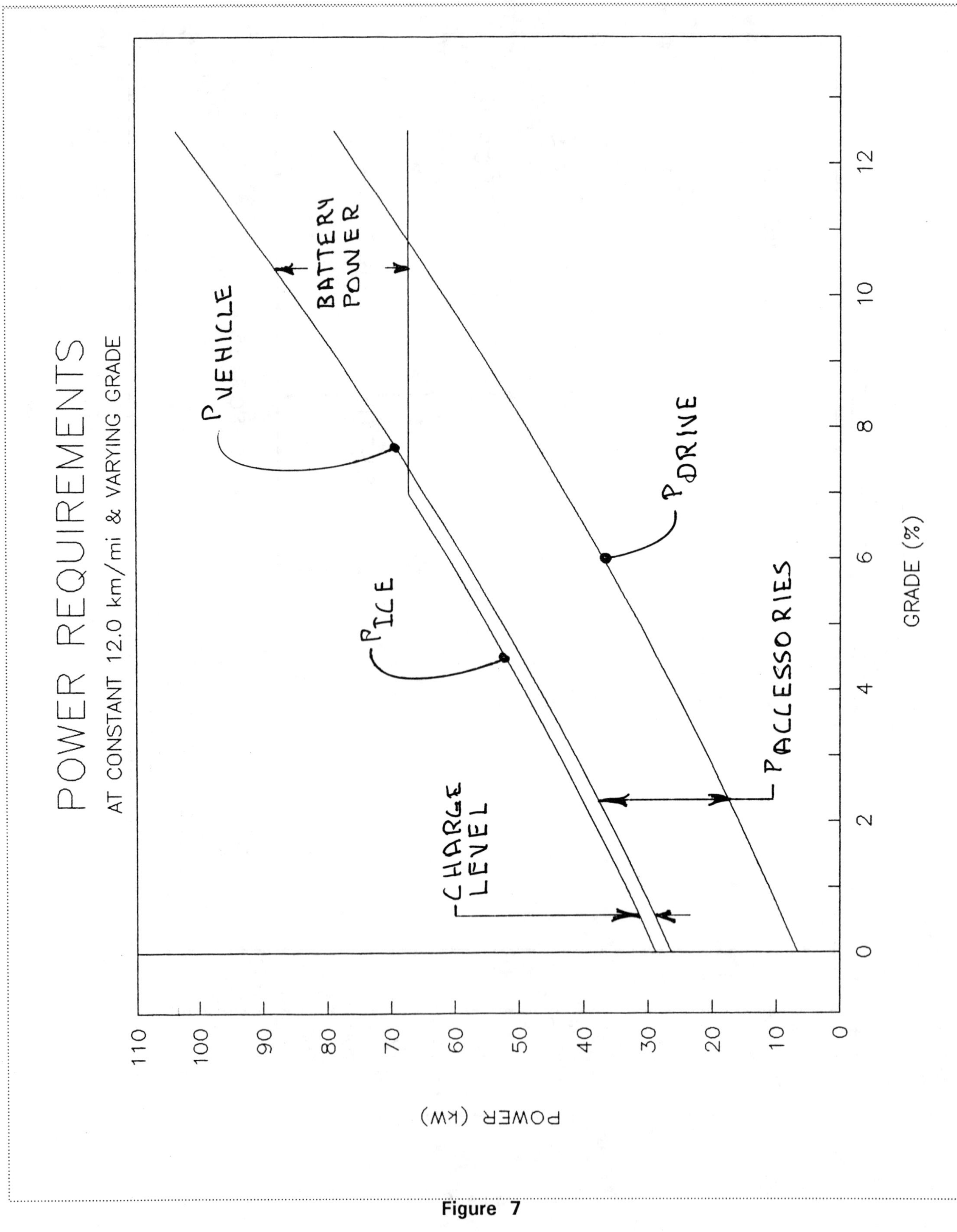
POWER REQUIREMENTS
AT CONSTANT 12.0 km/mi & VARYING GRADE
P VEHICLE
BATTERY POWER
P ICE
P DRIVE
CHARGE LEVEL
P ACCESSORIES
POWER (kW)
GRADE (%)
0
10
20
30
40
50
60
70
80
90
100
110
0
2
4
6
8
10
12

Figure 7

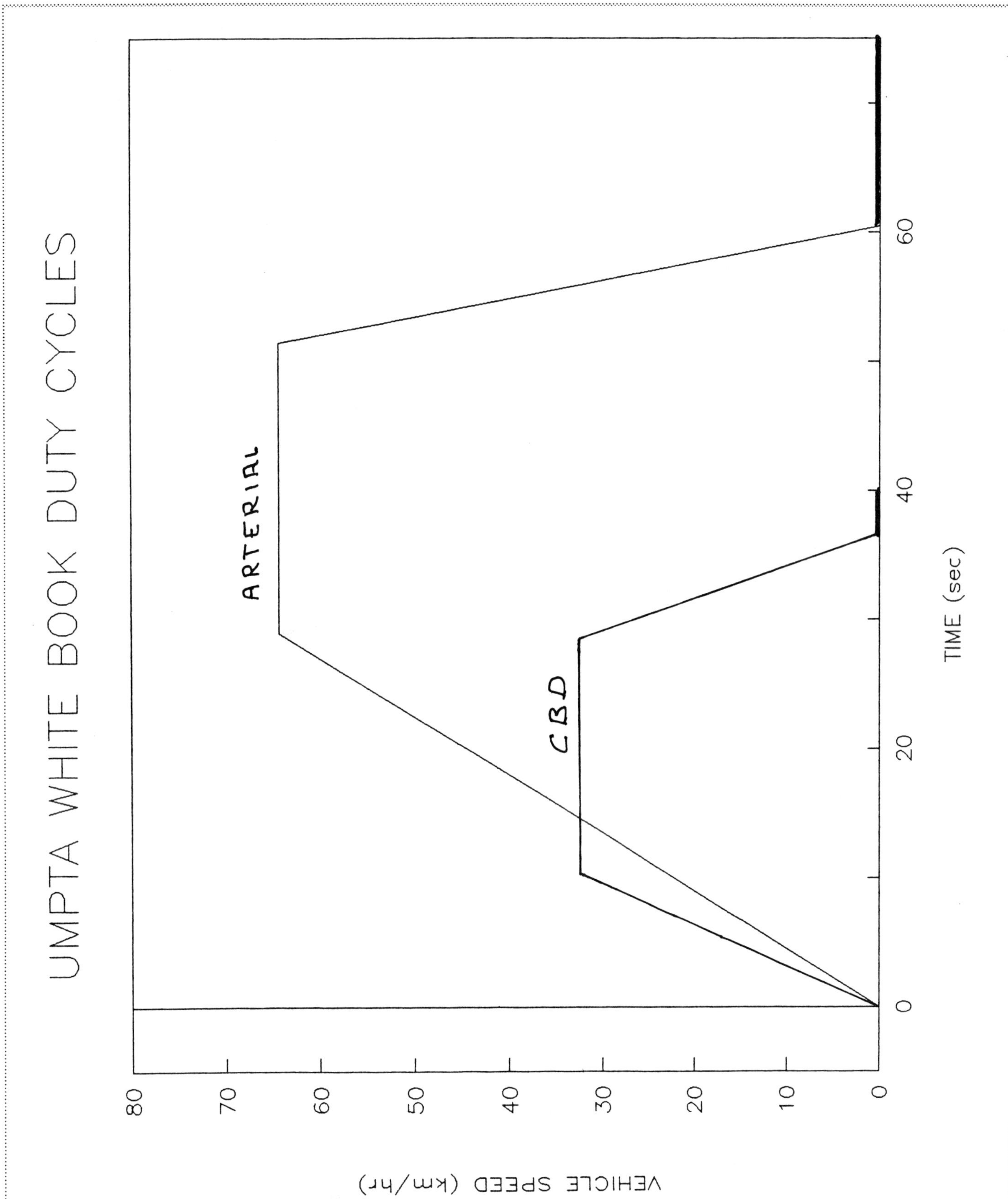
UMPTA WHITE BOOK DUTY CYCLES
ARTERIAL
CBD
VEHICLE SPEED (km/hr)
TIME (sec)
0
10
20
30
40
50
60
70
80
0
20
40
60

Figure 8

910249

Zinc-Air Powered Electric Vehicle Systems Integration Issues

Mike C. Cheiky and Len G. Danczyk
Dreisbach ElectroMotive, Inc.
Santa Barbara, CA

Robert L. Scheffler
Southern California Edison
Rosemead, CA

Abstract

Dreisbach ElectroMotive, Inc. (DEMI) is developing and testing a maintenance-free wall-plug rechargeable Zinc-Air Battery to power Electric Vehicles. This new battery technology offers over 200 mile vehicle range capability from very low cost, commonly available raw materials.

This paper will focus on the preliminary systems integration needed by an electric vehicle to operate these air breathing batteries in various common environments. Air cooling and reaction air requirements will be covered as well as the actual systems used to implement these requirements in a Chrysler minivan* which is sponsored by Southern California Edison (SCE). In addition, the projected system implementations of Zinc-Air batteries in automobiles and multi-use vans will be presented.

* Minivan refers to the generic T-115 chassis from Chrysler Corporation which is used on the Dodge Caravan and the Plymouth Voyager models.

** Numbers in parentheses designate references at end of paper.

Introduction

As reported at the SAE FTT '90 conference, DEMI has been developing rechargeable Zinc-Air traction batteries and testing them for Electric Vehicle (EV) propulsion under the sponsorship of SCE (1)**. This earlier paper also details this new battery's specifications, general characteristics and projected economics. Zinc-Air's chemistry is similar to an alkaline flashlight battery except for its oxidizer (oxygen) which is obtained from air. This enables the battery to have a very high energy density of up to 200 Wh/kg.

DEMI has made breakthroughs in air cathode construction and in making the overall battery rechargeable. The company holds several patents on this system (2).

The EV propulsion (Model 5) Zinc-Air battery cell measures 30.5 by 55.9 by 1.5 cm thick (12 by 22 by 0.6 inches) and weighs 2.5 kg (5.5 lbs). The cell has integral ribbing to support cell stacks of up to 25 cells high. Each cell features integral cathode air ducts that are separate and perpendicular to cooling air ducts for the cell above it.

As listed in Table 1, the Model 5 Zinc-Air energy density per weight is about 8X Lead-Acid, 4X Nickel-Iron, and 2X high temperature Sodium-Sulfur. The projected purchase cost of these Zinc-Air batteries in high volume production is $27. per kWh of battery capacity (1). This figure is less than one-half Lead-Acid's purchase cost (at $63./kWh), and much lower than other candidate EV battery technologies (1).

Battery Technology	Energy Density (in Wh/kg)	Electric Minivan Range with a 1000 lb (455 kg) Battery
Lead-Acid	30	30 mi (48 km)
Nickel-Iron	45	45 mi (72 km)
Sodium-Sulfur	90	90 mi (145 km)
DEMI's Zinc-Air	180	180 mi (290 km)

Table 1. Battery Technology versus Energy Density

Electric Vehicle Zinc-Air Testing

Model 5 Zinc-Air cells have been tested for one year in a Chrysler minivan which was converted to electric drive under SCE sponsorship. Various configurations of 48 cells for range and a comparable voltage Lead-Acid or Nickel-Cadmium battery for power have been tested and documented (1). This minivan was then upgraded to 84 and 168 Zinc-Air cell configurations and tested in urban and suburban drive cycles (3).

The best performance from testing as of November 1990 is 162 Wh/kg from the Zinc-Air cells observed in city driving (1). The best performance observed in mixed city and suburban driving is a total of 121.5 miles from 443 kg (975 lbs) of Zinc-Air batteries yielding about 140 Wh/kg of usable energy density (3). These initial tests also revealed efficiency improvement opportunities that are now incorporated in the parallel hybrid configuration below.

The initial tests utilized a series hybrid configuration whereby the Zinc-Air batteries constantly float charge the Lead-Acid acceleration batteries. This configuration is represented by the schematic in Figure 1 with the Lead-Acid recharge switch always closed. Such a configuration works satisfactorily in stop-and-go city driving apparently because the Lead-Acid Batteries are rapidly shallow cycled by being charged at every stoplight and discharged during acceleration. However, on longer suburban and freeway driving there are long periods between stops resulting in Lead-Acid efficiency losses of over 30 percent (3).

Initial testing as of November 1990 with the parallel hybrid configuration of Figure 1 eliminates most of this loss but requires the Zinc-Air batteries to have sufficient power output to support the expected cruise speed of the vehicle.

A total of 700 miles of in-vehicle testing have been made on the Zinc-Air cells as of November 1990. These results in conjunction with laboratory cycle tests indicate that up to 15,000 miles (24,000 km) of in-vehicle service life may be obtained from this type of Zinc-Air battery. The large scale production economics of this battery replacement interval along with energy costs result in a minivan operation cost (major maintenance and energy) of 16.4 cents per mile (1). In comparison to the gasoline powered version of this minivan (1989 Plymouth Voyager), this is equivalent to gasoline priced at $1.50 to $2. per gallon.

Since the energy consumption of a 2-passenger Sports Coupe or a 4-passenger Sedan can be lower per mile than a minivan, their Zinc-Air powered economics are projected to be close to today's gas-powered operating cost. For example, a 2-passenger "maximum range" Zinc-Air configured Sports Coupe consuming 125 Wh/mi (78 Wh/km) would need battery replacement every 26,000 miles (42,000 km) (1). Its combined energy and battery replacement cost is less than 8 cents per mile whereas the average 5-year operating cost (fuel and maintenance only) of 1989 "Sport" class vehicles range from 9.8 to 17 cents per mile based on assumed gasoline costs of $1.00 per gallon (4).

Configuration Details

As shown in Figure 1, the current minivan configuration is a parallel hybrid of 100 or 200 Zinc-Air cells for range and a 96-volt Lead-Acid battery (8 12-volt batteries) for power. Alternately, Nickel-Cadmium or super capacitors could be used in place of the Lead-Acid battery. This parallel hybrid configuration uses relatively high power Zinc-Air cells which provide all of the vehicle's energy when cruising at a steady speed. The power provided by the Lead-Acid battery is used for passing, acceleration, or a steep mountain incline. This configuration also lowers the power density needed by the Zinc-Air cells which allows them to last longer. In future configurations, the peak power-providing battery could also accept most of the energy from regenerative braking which was not implemented in the initial Zinc-Air powered test minivan to simplify battery testing.

This battery-battery hybrid configuration enables EVs that are comparable to gasoline powered vehicles in performance, cost, weight, and range of 200 to 300 miles (320 to 480 km). Battery recharging is not required every night and could be done as infrequently as a fuel tank fill-up in a gasoline powered vehicle. Alternately, in an average-use of 25 to 50 miles (40 to 80 km) per day with shorter recharges overnight to top-off capacity, the vehicle would always be ready for a longer continuous trip of 200 to 300 miles (320 to 480 km).

System Integration Issues

Zinc-Air to Lead-Acid Ratio

Zinc-Air cells are power limited in direct proportion to the Air Cathode surface area so near term configurations should be sized so they can provide all of the vehicle's cruise or constant-speed power. The Lead-Acid battery should be sized for the vehicle's desired acceleration 'duty cycle'. If the Lead-Acid battery is too small, it may easily fatigue (become discharged or develop high impedance) and, if it is too large, it will dilute the overall energy density of the vehicle. Therefore this ratio must be carefully chosen based on the vehicle's design goals of performance, gradeability, payload, durability, etc. In typical passenger vehicles such as the current test minivan, 400 to 600 lbs (180 to 270 kg) of Lead-Acid batteries appear to provide a workable solution.

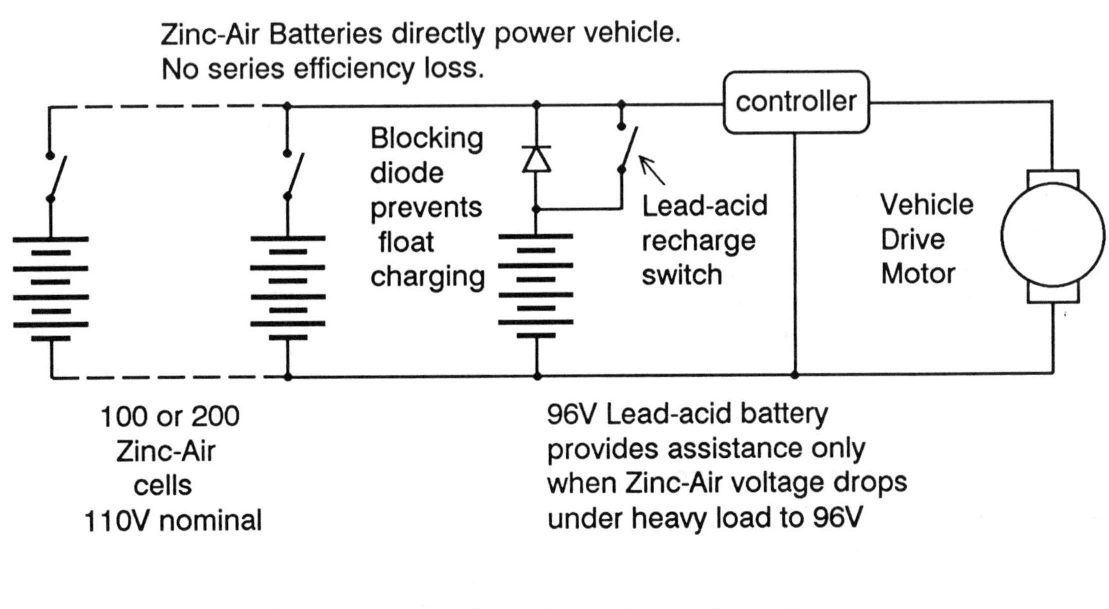

Figure 1. Parallel Hybrid Configuration

Air Cooling Needs

Since Zinc-Air cell's open circuit voltage is 1.4 volts and they deliver their rated current between 1.1 and 1.2 volts, they are about 80 percent efficient on discharge which corresponds to a 1.12 volt average per cell. The remaining 20 percent results in waste heat which internally heats the batteries. Cooling air is needed to keep the cells temperature from becoming more than 2°C above cathode air or above 25° C, whichever is higher, which tends to dry out the cells.

In minivan testing, an array of 8 3-inch electronic equipment fans each rated at 0.73 m^3/min. (25.7 cfm) maximum air flow are used per 25 cell stack. In the 100 cell configuration, this results in a maximum cooling capacity of 23 m^3/min (822 cfm). However, a production design could use a ram air duct from the frontal area of the vehicle to greatly assist air movement when it is most needed at higher cruising speeds.

Assuming an electric vehicle with 200 cells each rated at 40 amps, 2.24 kW of heat will be dissipated at rated power draw. The cooling air flow required to dissipate all of this waste heat with only a 2°C air temperature rise is calculated as follows. From a common automotive handbook, the change in Enthalpy, or heat content is equal to the Volume times Density times Specific Heat times the change in temperature. Or, in equation terms (5):

$$\Delta H = V * D * C * \Delta T$$

To obtain heat-per-unit-time terms of Joules per minute, divide both sides of the equation by time, t. To convert kW to kJ/min, a conversion factor and divisor of 60 to change hours to minutes is needed so that 2.24 kW equals 134.4 kJ/min (6):

$$2.24 \text{ kW} * (3.6*10^3) / 60 = 134.4 \text{ kJ/min}$$

In addition, since Volume-per-unit-time is needed, dividing both sides of the equation by D, C, and ΔT results in:

$$(\Delta H) / (t * D * C * \Delta T) = V / t$$

Plugging in values for the density of air, the constant volume specific heat value for air, and a temperature change of 2° K results in (5):

$$\frac{134.4 \text{ kJ}}{1 \text{ min} * 1.293 \text{ kg/m}^3 * 0.716 \text{ kJ/kg °K} * 2\text{°K}} = V/t$$

which calculates to 72.6 m^3/min, or, with an appropriate conversion factor is equal to 2,564 cfm (6). Assuming a margin of safety of 2X, 145.2 m^3/min or 5,128 cfm is needed. A square foot (.093 m^2) of vehicle frontal area traveling at 60 mph (97 km/h) or 5,280 feet per minute (1609 m/min) experiences an air flow rate of 5,280 cfm (150 m^3/min), which is less than many vehicle radiator grills in use today. Therefore Zinc-Air's battery cooling requirements are well within today's automotive experience.

When ambient temperatures are lower than 25°C, the cooling air should be throttled back to maintain at least a 25°C battery case temperature to allow generation of rated power. In very cold environments (-10°C and lower) it may be necessary to initially use the available battery power to preheat incoming air until the battery is up to power. In extreme cold, external preheat may be required, however a laboratory cell stored in liquid nitrogen (-196°C) achieved usable power in 2 minutes and full power in 10 minutes when fed room temperature air.

Carbon Dioxide Scrubbing from Reaction Air

Carbon dioxide (present in the air) needs to be removed from the reaction air because it may react with the KOH-based electrolyte in Zinc-Air batteries and slowly neutralize the electrolyte, thus limiting the batteries' performance and cycle life. Fortunately, carbon dioxide (CO_2) is a small constituent of air and the following calculations determine how much of it will be encountered.

The cathode or reaction air use rate at sea level and 22° C is only 13.6 L/min (0.48 cfm) per kW of power draw (1). For the test minivan traveling at 45 mph (72 km/h) and drawing 250 Wh/mi (155 Wh/km), this implies a 5-inch (13 cm) diameter electronic equipment cooling fan can supply the proper amount of reaction air. This fan limits reaction air

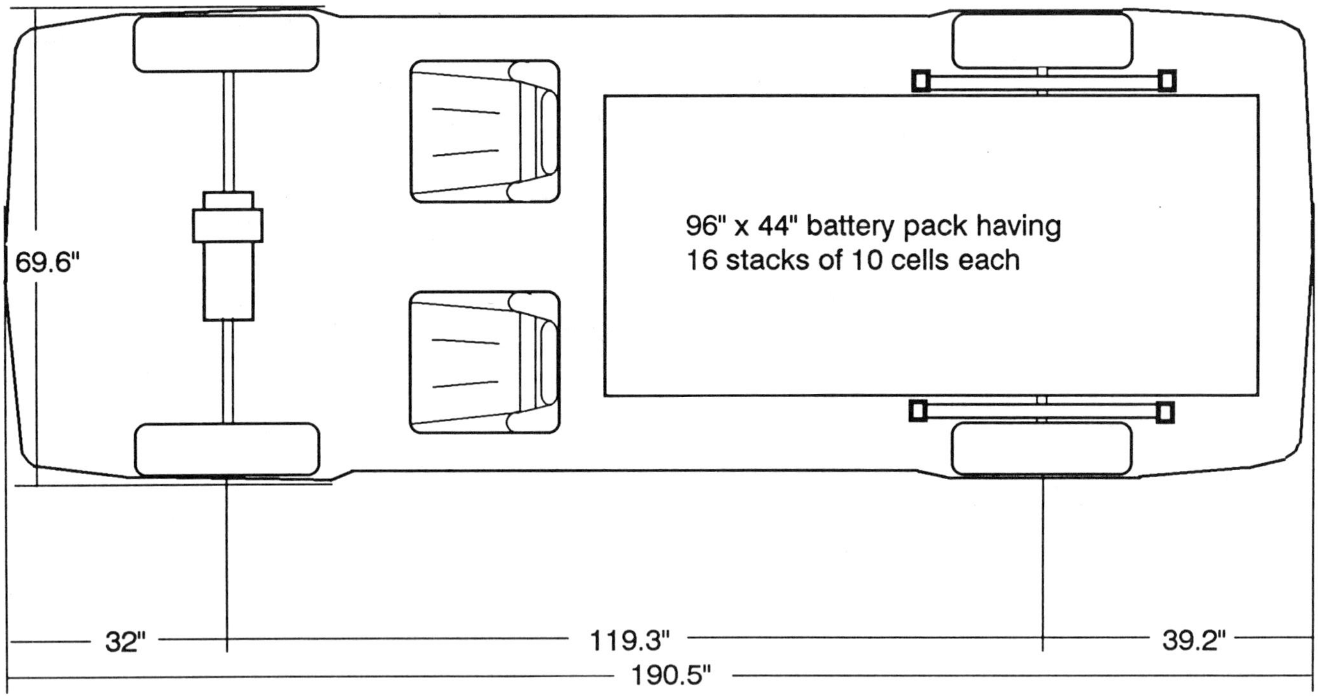

Scale = 28 : 1

**Figure 2. Minivan APV
Traction Battery Location**

flow to between 4 to 10 times the actual air being absorbed. This minimizes CO_2 absorption and the drying out or taking up of water by the cells from differing humidity levels in the air. The remaining CO_2 is scrubbed out of the reaction air by use of a heat reversible CO_2 absorber, lithium hydroxide (LiOH), which is added to an air filter to absorb CO_2 on discharge.

In order to build a filter that will last the life of the batteries, the amount of CO_2 exposure and the needed amount of LiOH is calculated as follows:

For a 200 cell vehicle example drawing 40 A @ 1.12 volts per cell, 8.96 kW of Zinc-Air power is provided. When traveling at 55 mph (88 km/h), a plausible 4-passenger Sedan with accessories has an estimated energy consumption of 163 Wh/mi (101 Wh/km). The battery's reaction air use at four times (4X) the stoichiometric (or actual use) rate is 54.4 L/min (1.92 cfm) per kW of power draw (1). Therefore, the 200 cell vehicle would need 487 L/min (17.2 cfm) of air (@ 8.96 kW draw).

The CO_2 content of air is equal to .033 percent (6). Multiplying this times the vehicle's air flow means the vehicle needs to scrub 0.16 L/min (.0057 cfm) of CO_2 from the air.

If a 5 hour (300 minute) per battery discharge is assumed (5 hours @ 55 mph (88 km/h) = 275 mile (442 km) range), a total of 48 liters (1.7 ft^3) of CO_2 must be removed. Expressed in molar terms by dividing by the conversion of 22.4 L/mole results in 2.14 moles of CO_2.

The conversion of CO_2 is by the following chemical equation where lithium hydroxide monohydrate is converted to lithium carbonate:

$$2LiOH * H_2O + CO_2 \rightarrow Li_2CO_3 + 2H_2O$$

Therefore a net of 84 grams of lithium hydroxide is needed per mole of CO_2.

Assuming a 15,000 miles (24,000 km) battery service life and 275 miles (442 km) per discharge results in 54 discharge cycles. Multiplying

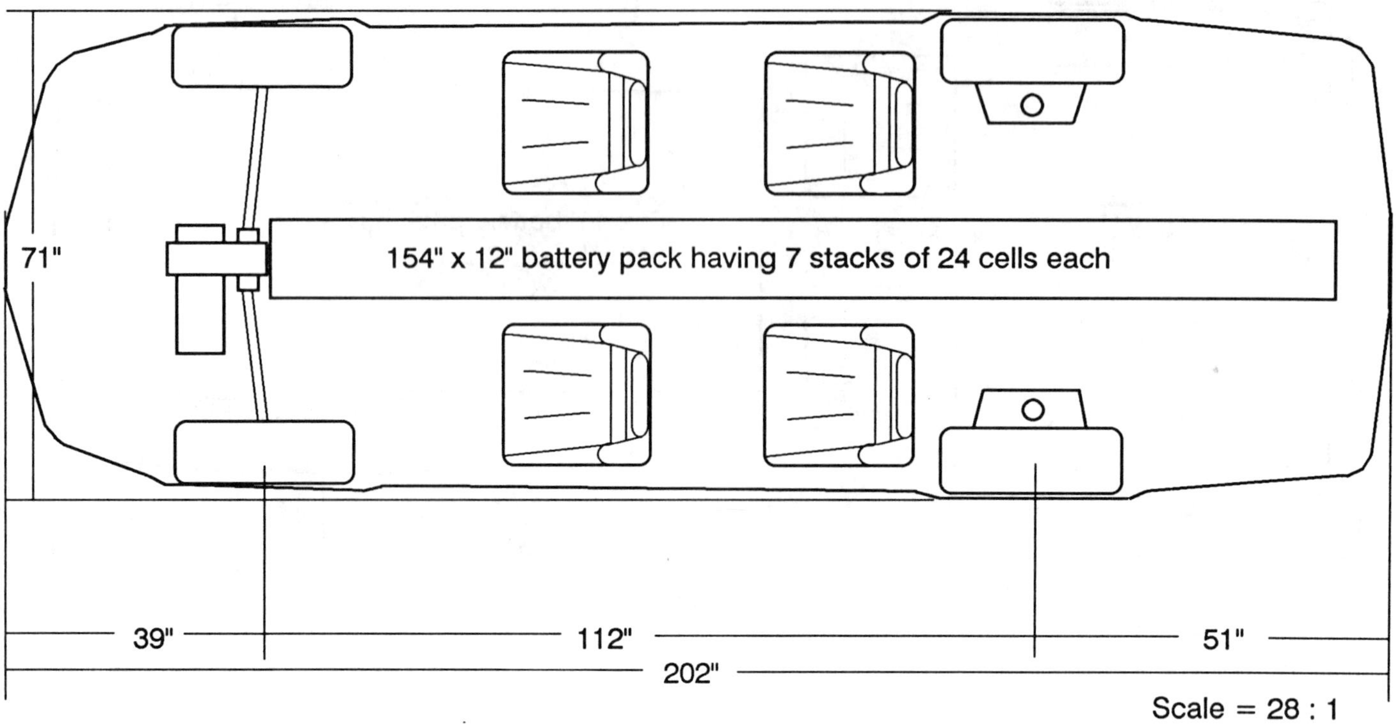

Figure 3. 4-Passenger Sedan Traction Battery Location

this by 2.14 moles of CO_2 per discharge from above and 84 grams of lithium hydroxide per mole of CO_2 results in a battery life requirement of less than 10 kg (22 lbs) of lithium hydroxide. The material cost of this compound amount is about $40., making it a feasible service item which can be recovered by high temperature heating. Alternately, it could be contained in a disposable filter or incorporated in the battery pack.

The area of CO_2 scrubbing has been extensively studied by NASA for manned spacecraft. Many alternate CO_2 scrubbing solutions are available including using other alkalies, amines, or zeolites.

Humidity Control Issues

The gelled electrolyte loses water when the cells are hotter than ambient. A Model 5 Zinc-Air cell has an equilibrium vapor pressure equivalent to 45 percent relative humidity at 25°C. To attain long cycle life, the input reaction air stream usually needs humidification during discharge. In rare cases when high humidity exists, simply operating the battery at a slightly higher temperature than ambient (less cooling energy needed) can easily compensate for increased moisture.

A typical environment that a vehicle must handle is when dry cold air is heated thereby requiring all of its water to be added to bring it to 45 percent relative humidity at 25°C. The partial pressure of water is 23.8 mm per 760 mm at sea level and at 25°C (6). Multiplying this ratio by 0.45 for the humidity factor and the 200 cell vehicle's 4X air flow rate of 487 L/min results in 6.86 L/min (.24 cfm) of water containing air at 55 mph (88 km/h). Using a per-discharge basis multiple of 300 minutes results in a typical water need of 2058 liters (72.7 ft^3) of humid air per discharge. Dividing by 22.4 L/mole and multiplying by 18 grams per mole results in 1.65 kg or 1.65 liters (.436 gal) of water needed per discharge. Therefore an 8 liter or 2-3 gallon vehicle water tank should be reasonable to hold enough for multiple discharges or well over one desert-travel discharge at high temperature (worst case). The

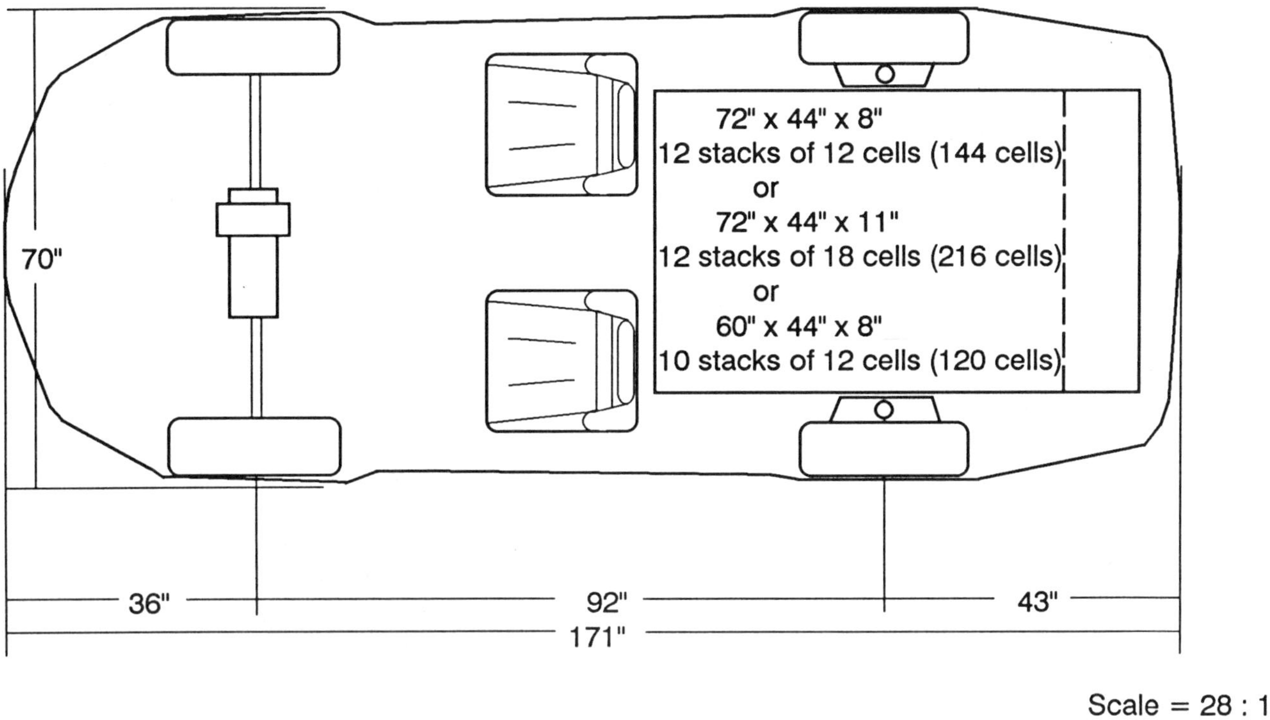

Figure 4. 2-Seat Sports Coupe Traction Battery Location

typical water consumption for operation in most of the continental United States is expected to be one to two gallons (4 to 8 liters) per month at a 12,000 mile (19,300 km) per year operating rate.

Projected Vehicle Configurations

Figure 2 shows the Zinc-Air battery placement in a minivan. This configuration accommodates a 200 mile (320 km) range with an additional 50 mile (80 km) limp home capacity which DEMI refers to as a 'reserve'. This configuration implements battery placement under the floorboard of a minivan such as the Chrysler T-115 chassis without major cargo space and height changes in comparison to the gasoline powered version of the minivan.

Figure 3 depicts a possible Zinc-Air battery placement for a 4-passenger Sedan with 250 mile (400 km) range and 50 miles (80 km) of reserve. The battery pack would fit in an enlarged transmission drive shaft tunnel as found in Sports class 2-door hardtops. The tunnel would be camouflaged with a center console extending to the rear seat. A fifth passenger position could be located between the seats but this location may not permit a comfortable seat.

The 2-passenger Sports Coupe Zinc-Air battery placement shown in Figure 4 has three options listed and many others possible. As shown it could be configured for high performance with a larger conventional battery for power and 144 Zinc-Air cells for range. Alternately, it could have only 120 Zinc-Air cells to maximize cargo space. Also possible is a total of 216 Zinc-Air cells to give the vehicle up to 300 miles (480 km) of range.

In all cases a slide-in battery module is anticipated which would be accessible by unlatching a hinged door or pivoting a hinged rear bumper assembly out of the way. In addition, also assumed and not shown in Figures 2-4 is an energy concentrator such as a small Lead-Acid pack which could be located over the front wheels of the vehicle.

Conclusion

The issues of integrating support systems for a Zinc-Air powered Electric Vehicle are readily addressed with solutions that are well within normal vehicle design contraints. The in-vehicle testing to date implies this technology can provide very economical long range Electric Vehicles that are also weight and cost competitive with gasoline powered vehicles. Providing that the cycle life of integrated Zinc-Air powered vehicles approaches the results of laboratory cycle life tests, these Zinc-Air cells can provide economically viable Electric Vehicle propulsion.

References

1. Cheiky, M.C., Danczyk, L.G., Wehrey, M.C., "Rechargeable Zinc-Air Batteries in Electric Vehicle Applications", SAE paper #901516, Future Transportation Technology Conference and Exposition, San Diego, CA, August 13-16, 1990.
2. Cheiky, M.C., "Metal-Alloy Air Battery", U.S. Patent #4,894,295, January 15, 1991; "Metal-Air Battery Power Supply", U.S. Patent #4,913,983, April 3, 1990; "Rechargeable Metal-Air Battery", U.S. Patent #4,957,826, September 18, 1990., Additional U.S. and Foreign Patents pending.
3. Cheiky, M.C., Danczyk, L.G., Scheffler, R.L., "Urban and Suburban Drive Cycle Test Results of a Zinc-Air Powered Vehicle", Proceedings of the 6th Annual Batttery Conference on Applications and Advances, California State University Long Beach, January 15-17, 1991.
4. Intellichoice, Inc., "Appendix B, Lowest/Highest Fuel and Maintenance Cost & Plymouth Voyager 3 Door Van", The Complete Car Cost Guide, 1989, pp 64A-65A & 145.
5. Adler, U. et al, "Heat, Enthalpy" and "Properties of Gases". Automotive Handbook 2nd Edition, 1986, Bosch, Robert GmbH, English Edition by SAE, Inc., Warrendale, PA, pp 84 & 174.
6. Weast, Robert C., "Conversion Factors", "Components of Atmospheric Air" & "Vapor Pressure of Water below 100°C", Handbook of Chemistry and Physics, 62nd Edition, 1981-1982, CRC Press, Boca Raton, FL, pp F-295, F-172 & D-168, resp.

910250

Climate Control for Electric Vehicles

John Dieckmann and David Mallory
Arthur D. Little, Inc.

ABSTRACT

The vast majority of cars and small trucks are sold with factory installed air conditioning (approximately 80% in 1989). For electric vehicles to succeed in the marketplace, air conditioning will need to be offered as optional equipment, along with adequate heating and defrosting systems. While providing the level of cooling performance expected by vehicle operators, it is important that the power consumption of the air conditioning systems used in electric vehicles be minimized, to minimize penalties to vehicle range and performance.

This paper summarizes the design and performance of several air conditioning systems that have been developed for electric vans over the past two years, including systems based largely on standard automobile air conditioning components and more advanced systems using high performance heat transfer components and a variable speed refrigerant compressor. The feasibility of using a heat pump cycle for vehicle heating functions (instead of a fuel fired heater), and the feasibility of applying measures to control vehicle thermal loads is discussed.

INTRODUCTION

As concerns about local levels of air pollution have risen, governments have legislated incentives and requirements for zero emission and low emission vehicles to make up a portion of each manufacturers fleet sales. The most notable example of this is legislation recently enacted in California requiring a mix of various levels of "low" emission vehicles, according to a formula, in each manufacturers fleet, beginning in 1998. Electric vehicles are well positioned to fill a significant portion of the market niche crated by this development, because the basic electric vehicle emits no pollutants. To capitalize fully on this opportunity will require that EV's be offered with sufficient driving range on a single battery charge to satisfy the requirements of a substantial portion of vehicle purchasers and that the zero or very low emission status be maintained to maximize the extent to which the low emission vehicle requirements are satisfied.

Vehicle climate control systems have a significant impact on both of these accounts. Air conditioning systems consume more power than any other accessory item; the power consumption must be minimized to minimize the impact on the vehicle range. Currently available liquid fuel fired heating systems operate at emission levels that result in forfeiture of zero emission status. To the extent that heating functions can be satisfied by waste heat recovery and heat pump cycle operation of the air conditioning system, the zero emission status of an EV can be maintained.

This paper addresses two major areas: the design and the performance test results of several EV air conditioning systems that have been developed by the authors and their associates over the past two years, and the results of a recently completed study of design measures to reduce vehicle thermal loads and of the feasibility of using a heat pump cycle for vehicle heating functions.

DESCRIPTION OF ELECTRIC VEHICLES AIR CONDITIONED AND TESTED TO DATE

Prototype air conditioning systems have been developed and tested for two separate electric G-Vans (referred to as G-Vans "A" and "B" in this paper) and one electric TE-Van. Both G-Vans are full size, window type GMC Vans with finished interiors, tinted side windows, and a clear glass windshield and rear windows. Both G-Vans are powered by lead acid battery packs. G-Van "A" was built by Cars and Concepts in 1987. G-Van "B" was manufactured by Vehma in 1989. The TE-Van is an electronically powered Chrysler "T115" van

(Plymouth Voyager/Dodge Caravan), with the body and interior configured similarly to the G-Vans, and is powered by a nickel-iron battery pack. It was built by Pentastar Electronics and McKee Engineering.

COOLING LOADS - The cooling loads were calculated for three basic configurations of each van - panel van with unfinished interior, panel van with partition, and window van. The cooling loads for the two test vehicles were estimated by calculating the individual contribution to the load for each of four categories: solar, conduction, ventilation and metabolic.

Automotive air conditioning systems are normally sized to provide adequate cooling at 110°F ambient conditions. The air conditioning system can operate in either full vent or recirculation mode. In the latter case, air recirculated from the interior plus a small amount of ventilation air is cooled by the evaporator and distributed back to the interior.

The assumed ambient condition is 110°F, 40% RH, with the vehicle interior cooled to 80°F, 50% RH.

Solar and conduction loads were calculated for each van at stationary and 35 mph conditions. The front of each van was assumed to be oriented toward the sun with the slope of the windshield perpendicular to incident solar radiation. This implies that the full solar load is transmitted through the windshield. Solar radiation was assumed to be 333 Btu/(hr ft^2) - equivalent to a clear summer day. The glass transmittance relative to solar radiation was 83% (clear glass).

The conduction load is dependent on exterior skin temperatures, as well as material layers in the body panels. The roof temperature was assumed to be 150°F with the vehicle stationary and 120°F at 35 mph. The assumed windshield temperature was 120°F when stationary and 110°F at 35 mph. All other skin temperatures are assumed to be at ambient temperature.

The window van is assumed to have a finished interior with an air gap between the exterior sheet metal surface and the interior wall. A natural convection heat transfer coefficient of 1 Btu/(hr ft^2°F) was assumed both for heat transfer across the air gap and for heat transfer between the interior surface and interior air. Carpeting on the floor was assumed to reduce heat conduction to negligible levels. The panel van was assumed to have an unfinished interior. The partition is located approximately one foot behind the driver and passenger seats, in the partitioned panel van.

The ventilation load in conventional air conditioning systems represents well over 50% of the steady state cooling loads. Reducing the ventilation air to minimum levels significantly decreases the air conditioning system cooling requirements. Given that the Federal Motor Vehicle Safety Standard has no requirement for ventilation air in automobiles, the ASHRAE Ventilation for Acceptable Indoor Air Quality Code, ANSI/ASHRAE 62-1981R, was used as a baseline. The ASHRAE Code recommends 15 cfm of ventilation air per person for transportation vehicles.

The ventilation air for a panel van, which would have one driver and one passenger, would be 30 cfm. The ventilation air for a window van would be increased to 45 cfm, allowing additional fresh air for the extra passengers.

The estimated cooling loads with reduced ventilation air are shown in Tables 1 and 2. Calculation of solar, conduction, and metabolic loads are based on the same assumptions as those used in the conventional air conditioning system load calculations.

Table 1: G-Van Estimated Steady State Cooling Loads (Btu/hr)

				Ventilation Load		Metabolic Load Occupants		
Oper. Cond.	Van Type	Solar Load 83% Trans.	Conduction Load	30 cfm	45 cfm	2	6	Total
Stationary	Panel	3,450	12,180	2,490		1,000		19,120
	Panel with Partition	3,450	4,390	2,490		1,000		11,330
	Window	3,450	5,060		3,730		3,000	15,240
35 mph	Panel	3,450	10,620	2,490		1,000		17,560
	Panel w/Partition	3,450	3,550	2,490		1,000		10,490
	Window	3,450	3,880		3,730		3,000	14,060

Table 2: TE-Van Estimated Steady State Cooling Loads (Btu/hr)

				Ventilation Load		Metabolic Load Occupants		
Operating Cond.	Van Type	Solar Load 83% Trans.	Conduction Load	30 cfm	45 cfm	2	6	Total
Stationary	Panel	3,230	10,280	2,490		1,000		17,000
	Panel with Partition	3,230	4,340	2,490		1,000		11,060
	Window	3,230	4,900		3,730		3,000	14,860
35 mph	Panel	3,230	8,370	2,490		1,000		15,090
	Panel w/Partition	3,230	3,510	2,490		1,000		10,230
	Window	3,230	3,885		3,730		3,000	13,845

AIR CONDITIONING SYSTEM DESCRIPTION

Figures 1, 2, and 3 are schematics of the prototype systems that were installed in the three vehicles. The major components of each system are indicated in the schematic. The air conditioning system in G-Van "A" used a Harrison V-5 variable displacement compressor belt driven by a brush type permanent magnet DC motor. The compressor operating speed of 1,650 RPM was set to provide a maximum cooling capacity approximately 35% greater than the calculated steady state cooling load at a 110°F ambient, with the excess to provide for cool down from hot soak. The TE-Van system used a Sanden TRF 105 compressor, belt driven by a brush type permanent magnet DC motor. The motor/compressor assembly was located near the rear of the vehicle, the only available location within the prototype vehicle, with resulting, longer than normal compressor suction and discharge lines. The compressor operating speed was 1,200 RPM, set to provide enough capacity to meet the calculated steady state load at a 110°F ambient with a full solar load. In G-Van "B" a Sanden TRF 105 compressor belt driven by a variable speed electronically commutated motor was used. The motor and drive were manufactured by GE Motors. The maximum compressor speed of 3,000 RPM was set to provide rapid cool down, with the variable speed drive allowing the capacity to be matched to the load efficiently at steady state. The speed is controlled on the basis of meeting a user adjustable air delivery temperature set point.

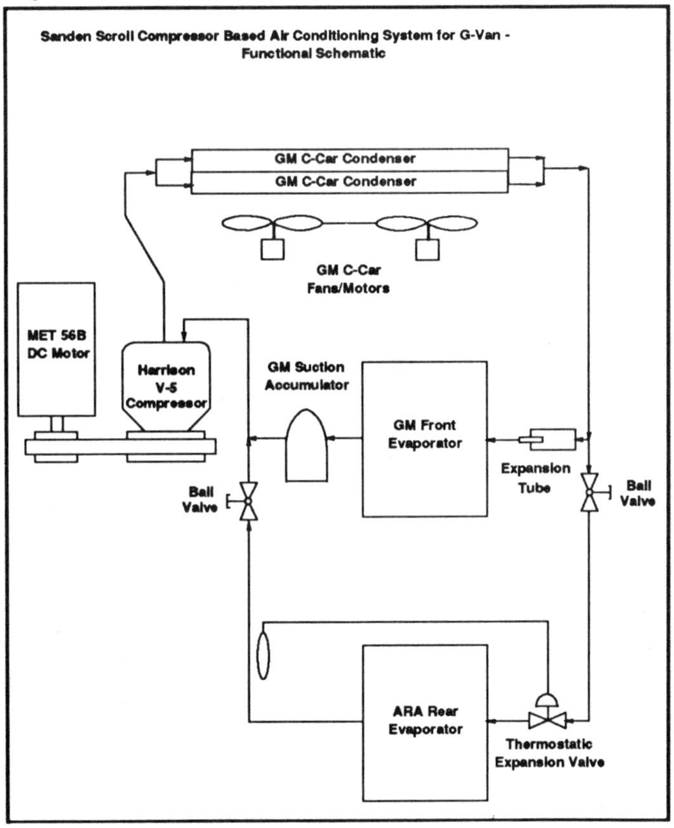

Figure 1: Refrigerant Flow (Schematic) for Constant Speed - Variable Displacement - Belt Driven Air Conditioning System Used in G-Van "A"

The standard front evaporator and air handler was used in each system. In the G-Van with the constant speed variable displacement compressor system, a rear evaporator/blower unit was used as well. To improve system efficiencies, higher capacity condensers were used in both G-Van systems. In the first G-Van, two Cadillac C-Car condensers, were installed in series on the air side, each condenser having 25% more frontal area than a standard G-Van condenser. In the second G-Van, a parallel flow, flat tube condenser (Modine PF) was used.

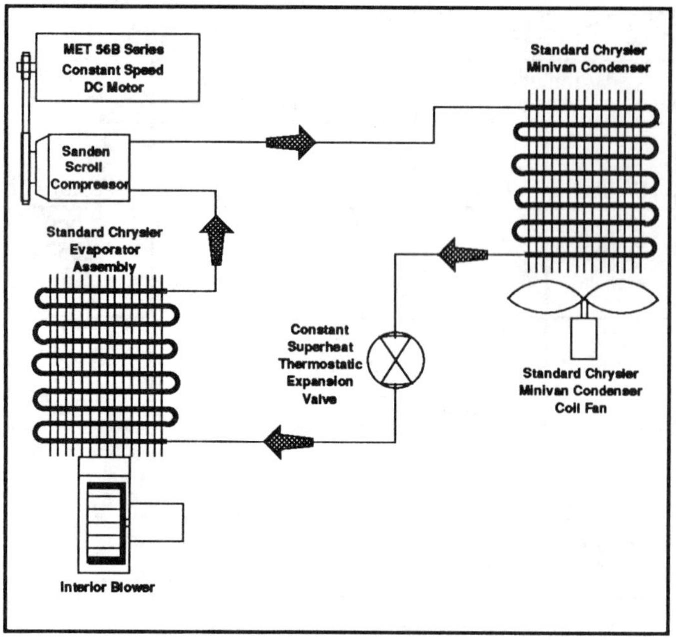

Figure 2: Refrigerant Flow Schmatic for Constant Speed Belt Driven Compressor Based System Used in TE Van

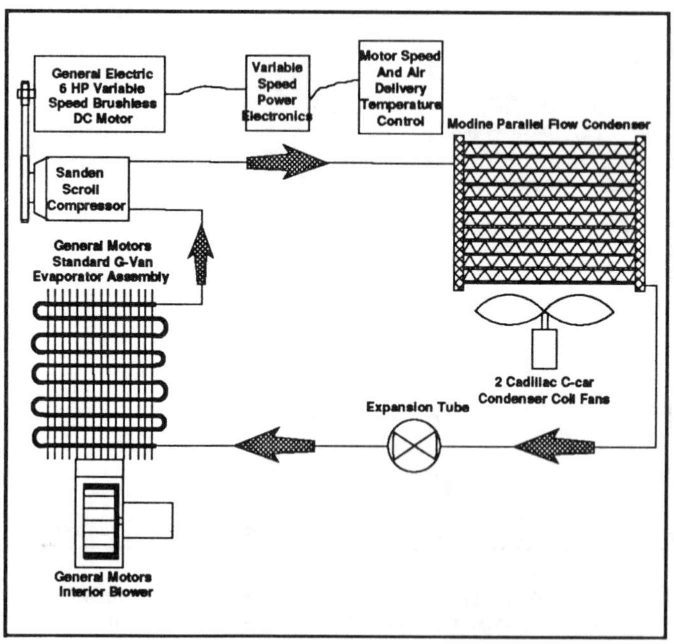

Figure 3: Refrigerant Flow Schematic for Variable Speed Compressor Based Air Conditioning System for G-Van "B"

The system weight break down is given in Table 3. Parasitic power consumption is summarized in Table 4.

Table 3: System Weight Breakdown

	Weights, lb		
Component	TE Van Constant Speed	G-Van Constant Speed	G-Van Variable Speed
Compressor	14	14	14
Compressor Motor	40	40	27
Motor Controller	5	5	10
Condenser	6	15	9
Condenser Fan Assembly	5	8	8
Evaporator	5	20 (incl rear)	5
Refrigerant Tubing and Hoses	4	8	8
Refrigerant Charge	2	5	2
Total	81	115	83

Table 4: Parasitic Power

	Parasitic Power Watts		
Component	TE-Van Constant Speed	G-Van Constant Speed	G-Van Variable Speed
Compressor Clutch	33	33	33
Motor Control	6	8	20
Condenser Fan	80	65	65
Front Evaporator Blower (maximum speed)	185	300	300
Rear Evaporator Blower (maximum speed)	---	215	---
Total	304	621	418

PERFORMANCE TEST RESULTS

Each of the three systems was performance tested. The G-Van "A" System was road tested outdoors on a cloudless day, at an ambient of 100°F. The other two systems were wind tunnel tested, at ambient temperatures of 110°F and 90°F, under a full solar load.

Figures 4, 5, 6 are each plots of the vehicle cool down from hot soak at the highest ambient temperature in which it was tested, and under full solar load. Table 5 summarizes the steady state power consumption of each air conditioning system at 30 MPH.

The air conditioning system in the TE-Van cooled the vehicle marginally at a 110°F ambient -- the cool down rate was slower than for a normal production T115 Van, and the steady state temperature that was maintained, 84°F, is 5°F higher than for a production T115 Van. At a 90°F ambient, comfortable vehicle interior conditions were readily provided, with sufficient excess capacity that the compressor cycled on and off while maintaining a driver breath temperature of 75°F. In the interest of minimizing air conditioning power consumption, the compressor speed/capacity had been set to meet the calculated steady state load at the 110°F ambient condition. The short fall in steady state cooling capacity could be compensated for by increasing the compressor speed from 1,200 to 1,500 RPM, but with an increase in the power consumption. The primary factors limiting overall system performance are the limited condenser frontal area and resulting high heat rejection temperature and the long run of refrigerant suction line to the rear location of the compressor.

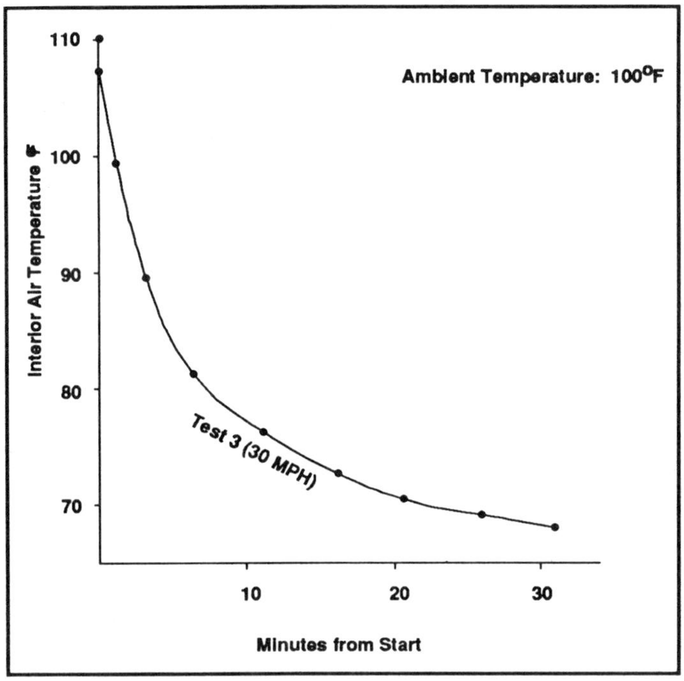

Figure 4: Cool Down from Hot Soak G-Van "A" (Front and Rear Evaporator Fans both at Maximum Speed)

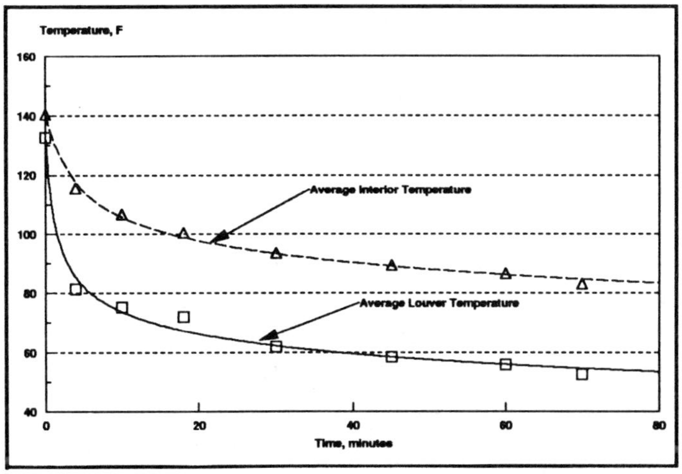

Figure 5: TE Van Cool Down Characteristics

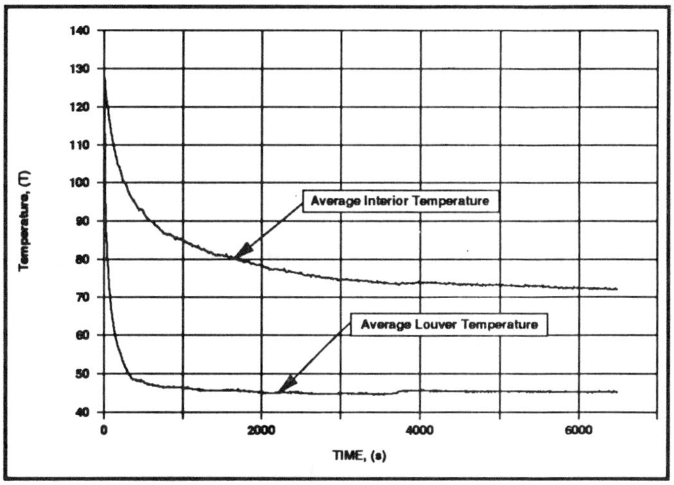

Figure 6: Variable Speed G-Van Cool Down Curve

The cooling performance of the two G-Van systems, both in terms of cool down rate and maintenance of steady state interior temperatures, was comparable to normal IC engine vehicle cooling systems. The power consumption levels shown in Table 5 result in the calculated range penalties shown in Figure 7. The reduced power consumption and improvement in range penalty of the variable speed scroll G-Van, "B" system is attributable to the higher efficiency of the Scroll compressor and of the electronically commutated motor, and the improved performance of the parallel flow condenser. At moderate ambients, the variable speed drive of the compressor and adjustable air outlet temperature allows the system output to be matched efficiently to the vehicle cooling requirements.

Table 5: Steady State Cooling Performance of the Prototype Air Conditioning Systems

				Electric Power Input Watt	
System	Ambient Temp °F/RH	Driver Breath Temp °F	Lower Temp °F	Compressor-Motor	Total
G-Van/Const. Speed Var Displacement Reciprocating	100 10% RH	70	42	2,150	2,400
TE Van/Const Speed Scroll	110 20%	84	58	2,200	2,450
	90 50 RH	72	43	1,400	1,600
G-Van/Variable Speed Scroll	110 40%	76	46	2,950	3,230
	90 50%	74	54	685	920

ONGOING EV AIR CONDITIONING DEVELOPMENT EFFORTS

In an ongoing program, two additional prototype EV air conditioning systems are under development. Both systems utilize a prototype semi hermetic scroll/variable speed electronically commutated motor/compressor and a parallel flow plate-fin condenser. The first of these systems has been installed in a prototype electric TE-Van and will be wind tunnel tested during the first quarter of 1991. The second system is designed for a Ford Aerostar based ETX-II prototype vehicle and will be tested in the second quarter of 1991. The results of these tests will be the subject of a future publication.

FUTURE EV CLIMATE CONTROL

VEHICLE THERMAL DESIGN - Reduction of the vehicle heating and cooling loads directly reduces the power consumption of vehicle climate control systems by reducing the heating or cooling that must be supplied by these systems. In addition, reduced capacity requirements can allow for either increased system efficiency within the constraint of fixed heat exchanger face areas and ducting air flow cross sectional areas, or downsizing of components for reduced weight and cost.

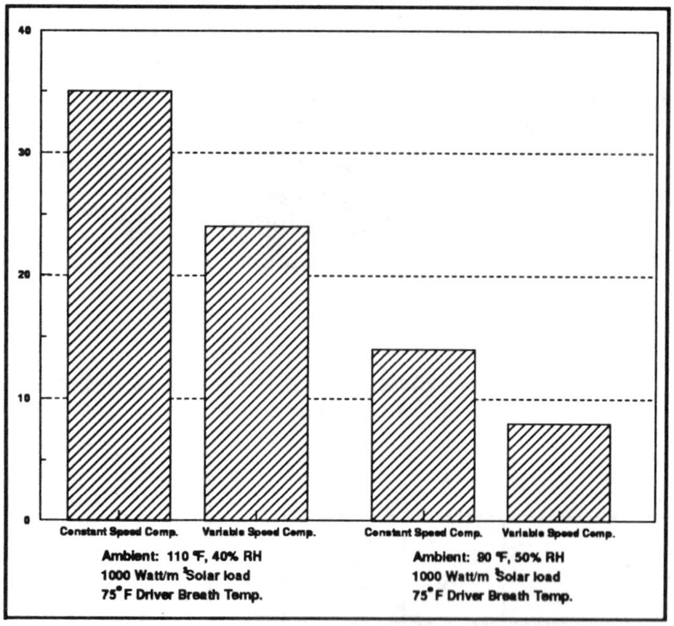

Figure 7: Air Conditioniong Range Penalties Prototype G-Van (SAE J227A C-Cycle)

To allow for sizing of air conditioning systems for electric vehicles and assessing options for reducing thermal loads, a PC based electric vehicle thermal model was developed, which was used in this study to assess a basic set of thermal load reduction measures for a typical mini-van (window van configuration).

Table 6 lists the design ambient temperature extremes for the heating and cooling season for a set of cities representing the range of climatic extremes in the continental United States (lower 48 States), based on the ASHRAE Fundamentals Handbook. Two sets of values are provided: the temperatures that are exceeded less than 1% and 5% of the cooling season, and the temperatures which are equal to or exceeded 99% and 97.5% of the heating season.

A vehicle thermal design configuration was defined, incorporating simple, low cost, effective measures to reduce steady state heating and cooling loads and hot soak temperatures. The thermal load model was used to compute the heating and cooling loads for configuration at the 1% design cooling temperatures and the 99% design heating temperatures. The basic thermal features are:

- Insulated roof (1/2 in thick rigid polyurethane);
- All glazings are wavelength selective, e.g. PPG Sungate (typical automotive glass characteristics are summarized in Table 7);
- Recirculation mode only (45 CFM of ventilation make up air); and
- 100 CFM of ventilation air, provided by operating the HVAC blower at approximately 5 volts and 20 watts (a reasonably efficient 20 watt DC/DC converter is required), is used to limit hot soak temperatures.

Figure 8 compares the cooling loads for a typical minivan incorporating the above design, with a typical production configuration having the ventilation air flow rate restricted to 45 CFM.

Significant reductions in both the calculated heating and cooling loads result, both at steady state and during cool down or heat up transients.

Table 6: Design Ambient Conditions For Six Climatic Regions In The United States

(Source: ASHRAE Fundamentals Handbook)

City	Design Heating Dry Bulb, °F		1% Design Cooling Temperatures, °F		5% Design Cooling Temperatures, °F	
	99%	97.5%	Dry Bulb	Wet Bulb	Dry Bulb	Wet Bulb
Phoenix	31	34	109	76	105	75
Los Angeles	37	40	93	72	86	70
Detroit	3	6	91	76	86	73
Minneapolis	-16	-12	92	77	86	73
Houston	28	33	97	80	93	79
Washington, D.C.	14	17	93	78	89	76

Table 7: Typical Automotive Glass Characteristics

		Entire Wavelength		
Glass Type	Visible Transmissivity	Transmissivity	Absorptivity	Reflectance
Ordinary Glass	.90	.84	.08	.08
Infrared Absorption Glass	.76	.53	.41	.06
Infrared Reflective Glass*	.73	.52	.11	.37
		Solar Wavelength		
Airco's Solar Control Film	.70	.34	.21?	.45
Adjustable Transparency Glass decolored colored	 .60 .05	 .52 .06		
Aircool Tinted windshield	.70	.38	.24	.38
Standard Tinted Windshield	.79	.57	.37	.06
PPG Sungate Glass	.71	.39	.30	.31

* Used in buildings, has not been examined in practical terms for motor vehicles

Source: SAE paper numbers: 880050, 861362, 890311, and 890217.

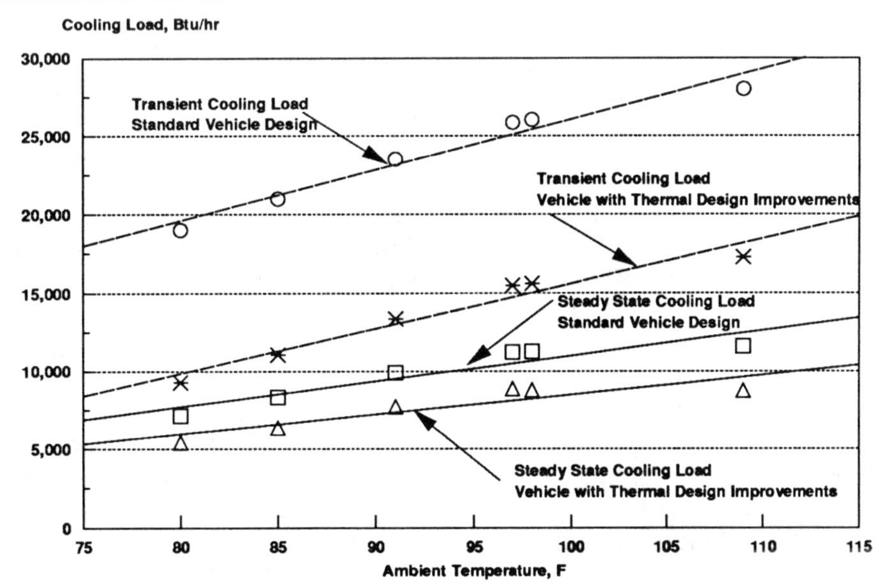

Figure 8: Typical Electric Mini-Van Cooling Loads

Table 8: Estimated Rate of Power Dissipation in the Train Controller and Traction Motor/Transmission for Selected Electric Vehicles

Vehicle	Driving Cycle*	DC Energy Consumption kWh/mi		Average Power**	Waste Heat Watts	
		Net	Gross	kW	Controller	Mot/Trns
Impact	FUDS 55	.110 .113	~.12 .113	2.6 6.2	210 125	490 600
EV Cort	FUDS 55	.341 .257	.364 .275	7.8 14.1	620 280	1430 1380
ETV-I	FUDS 55	.280 ~.20	~.31 .20	6.8 11.0	540 220	1250 1080
ETX-I	FUDS 55	.335 .260	.369 .260	8.1 14.3	650 285	1490 1400
ETX-II	FUDS 55	~.4 ~.4	~.43 ~.4	9.2 22	740 440	1700 2150
TE-Van	FUDS 55	~.4 ~.4	~.43 ~.4	9.2 22	740 400	1700 2150
G-Van	FUDS 55	.62 .63	.67 .63	14.6 35	1170 700	2700 3400

* Average speed in FUDS is 20 mi/hr; 55 means constant 55 mi/hr (88 km/hr)

** Total of traction and regeneration power through controller and motor/transmission

Table 9: Estimated Range Penalties For Vehicle Heating Using Vehicle Waste Heat And Heat Pump (Including Parasitics)

	55 mi/hr	FUDS*	FUDS/ Soaks**
Los Angeles Design Ambient (37°F)	1/2%	1%	5%
Los Angeles Heating Season Average Ambient (50°F)	1/4%	1/2%	3%
Washington, DC Design Ambient (14°F)	1%	4%	17%
Detroit Design Ambient (3°F)	2%	5%	22%
Minneapolis Design Ambient (-16°F)	4%	10%	36%

* Steady state, single trip, no initial cold soak, warmup time for inverter heat sink and motor/transmission neglected.

**Continuous series of 25 mile FUDS trips with full cold soak in between, no waste heat available from inverter or motor/transmission during first 20 minutes of each trip.

ELECTRIC VEHICLE HEATING - Most electric vehicles now use liquid fuel fired heaters to perform space heating and defrost functions. Although this approach is adequate in the near term for initial commercialization efforts, this arrangement has a number of short comings:

- The size and weight penalty associated with the fuel fired system is significant;
- The cost of the fuel fired heating system is significant;
- The need to service two energy sources on the vehicle will reduce the acceptability of electric vehicles to consumers;
- The exhaust from the heater will include NO_x, CO, hydrocarbons, etc., compromising the non-pulling (at least will respect to local air pollution) status of electric vehicles; and
- The fuel consumption of the heater can be significant, noticeably reducing the overall energy efficiency of the vehicle.

An alternative to the fuel fired heater is to use the air conditioning system as an electrically driven heat pump system. Heating would be provided by a combination of vehicle waste heat, the heat pump, and, when necessary, back up electric resistance heat. Such a system would simplify the space conditioning system design and reduce the overall size and weight of the space conditioning system by eliminating the need for a separate liquid fuel fired heater. It would also eliminate the need to service two energy sources on the vehicle thereby increasing consumer acceptance.

To illustrate the feasibility of this approach to vehicle heating, the minivan with improved thermal design example, above, was considered. This vehicle was assumed to have a heat pump whose maximum capacity is set to meet 150% of the steady state cooling load at the Phoenix design ambient. Recovery of heat dissipated in the traction motor inverter and the traction motor/transmission is assumed (Table 9 is a summary of the heat dissipation levels in these components for various electric vehicles). Figures 9 and 10 plot, versus ambient temperature, the sources of vehicle heat at steady state and during a warm up from hot soak (20 minute average). In general, recovered drive line waste heat and the heat pumps have sufficient capacity to meet the heating load at design ambients throughout the Continental U.S. Large amounts of back up electric resistance heat would be needed to warm the vehicle in a reasonable period of time in colder climates, but in areas having mild winters, the heat pump provides sufficient capacity for warmup, as well as steady state operation. Table 10 summarizes the estimated range penalties, for this heating technique, for three driving cycles. They are quite modest, except for a series of short trips in a colder climate.

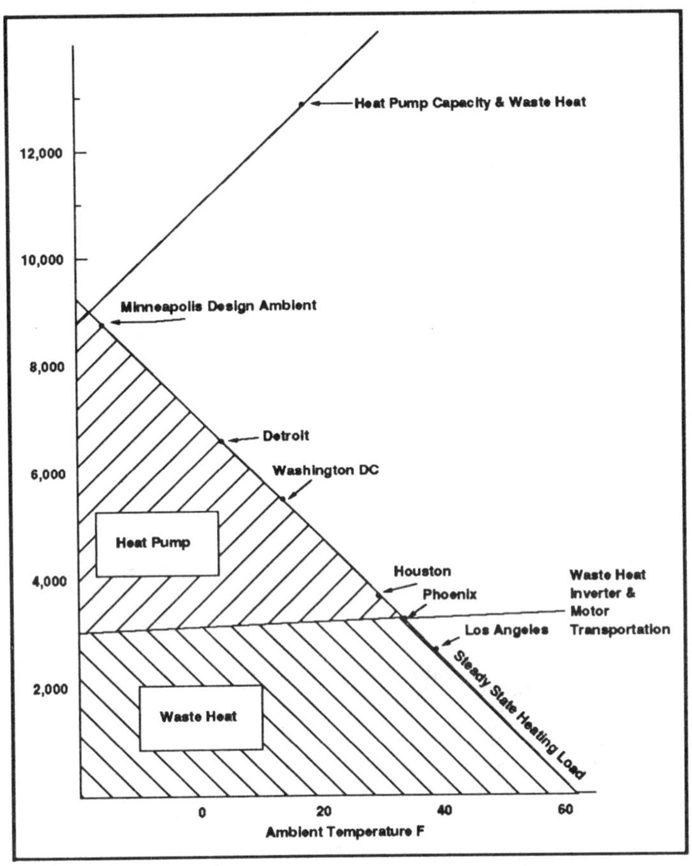

Figure 9: Steady State Heating of Baseline Minivan, Using Vehicle Waste Heat, Vapor Cycle Heat Pump and Back Up Resistance Heat to the Extent Necessary

Beyond capacity and power consumption, two other practical issues must be considered relative to application of a heat pump to vehicle heating:

- Defrosting the ambient air heat exchanger: in the heating mode, when the ambient air coil fin surface is below freezing, frost will accumulate, gradually building up and blocking airflow through the coil. In stationary heat pumps, the coil is defrosted by periodically reversing to cooling mode, but with back up heat supplied to the

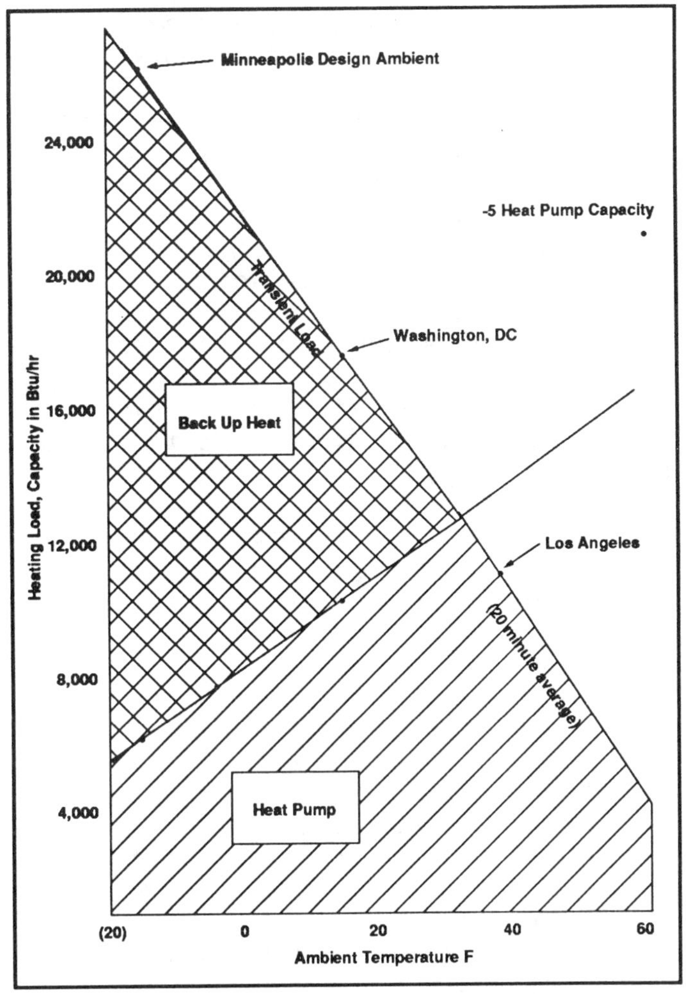

Figure 10: Transient Heating of Baseline Minivan, Using Vehicle Waste Heat, Vapor Cycle Heat Pump and Back Up Resistance Heat to the Extent Necessary

interior and the outdoor fan off. The outdoor coil is heated, melting the frost, then normal heat pump operation is resumed. In a moving vehicle, in cold weather, a means of blocking the air flow (e.g. louvers) during the defrost cycle may be needed to allow the coil to warm up quickly and melt the frost.
- Windshield defrosting: FMVSS 103 specifies minimum deicing performance, as measured by a specific test procedure. This requirement actually tends to drive the heating capacity requirement in some EV's. An electrically heated windshield (e.g. the Ford "Installation") is preferable.

CONCLUSIONS

Based on the performance of the prototype air conditioning systems developed to date, it is clear that air conditioning systems can be designed for electric vehicles that will meet normal expectations of cooling performance level, and have acceptable range penalties at extreme high temperature (110°F) ambient conditions and quite modest range penalties at average seasonal cooling conditions (85 - 90°F). Moreover, these performance levels have been achieved without resorting to any particularly "heroic" measures in either vehicle or air conditioning system design. The key elements required are limiting ventilation make up air flow rates to the levels normally required for human comfort (10 to 15 CFM per person), improving condenser performance levels over standard fin-tube condenser, and utilizing an efficient, continuously variable speed motor/compressor system.

Meaningful improvements can be achieved by consideration of vehicle thermal performance and applying a relatively modest set of thermal load reduction measures in addition to limiting ventilation make up air flow rates - wave length selective glazing, roof and body panel insulation and a low flow rate ventilation operating mode of the interior blower to limit hot soak temperatures.

Initial heat pump feasibility results show that for milder climates, such as Southern California, vehicle heating functions including rapid warm up, steady state heating, and window defrosting/defogging -- can be readily provided by a combination of recoverable vehicle waste heat sources, heat pump operation, and very limited amounts of electric resistance heat, at a very modest vehicle range penalty. Demonstration projects are needed in the near term to establish the feasibility for late 1990's production.

ACKNOWLEDGEMENTS

The development of the prototype electric vehicle air conditioning systems described herein was carried out under EPRI contract RP-2861-3. This project was cofunded by EPRI, Southern California Edison Company, and Pacific Gas and Electric Company. The study of electric vehicle thermal design options and heat pump feasibility was carried out under INEL subcontract No. C89-102531 under DOE Contract No. DE-AC07-76IDO1570. The support of these organizations is acknowledged.

REFERENCES

1. Dieckmann, J.T. and Teagan, W.P., Air Conditioning Systems for Electric Vehicles Phase I/II Final Report, prepared by Arthur D. Little, Inc. under EPRI Research Project 2861-3, May, 1990.
2. Crumly, R.L., et.al., Performance Testing and System Evaluation of the Soleq Evcort Electric Vehicle. EG&G Idaho, Inc. DOE/ID-10232. March, 1989.
3. Crumly, R.L., et.al., Vehicle Performance Tests of the Ford/GE First Generation Single-Shaft (ETX-I) Alternating Current Propulsion System. EG&G Idaho, Inc. DOE/ID-10224. April, 1989.
4. Burke, A.F., Comparison of Simulation and Test for Electric Vehicles of Recent Design. EG&G Idaho, Inc. SAE-890818. March, 1989.
5. Burke, A.F., Electric Vehicle Design and Performance Using Advanced Batteries. EG&G Idaho, Inc. SAE Technical Paper Series #891663, August, 1989.
6. Electric Vehicle Test Procedure, SAE Recommended Practice, J227a.
7. General Motors Corporation, General Motors Technical Center. Press Release dated January 3, 1990, describing the "Impact" prototype electric vehicle.
8. Kawamoto, H., et.al., Comfort Evaluation of Heating and Air Conditioning Systems, SAE Technical Paper Series #860590, February, 1986.
9. Bessler, W., and Forbes, F., Electrically Driven Automotive Air Conditioner, SAE Technical Paper Series #870035, February, 1987.
10. Goodremote, C.E., et.al., Compact Air Cooled Air Conditioning Condenser, SAE Technical Paper Series #880445, March, 1988.
11. Ikeda, S., Yoshii, Y., Tamura, Y., Air Conditioning Electric Vehicles with an Electronically Driven Variable Speed Scroll Type Compressor, SAE Technical Paper Series #901738, September, 1990.
12. Sullivan, R., Selkowitz, S., Effects of Glazing and Ventilation Options on Automobile Air Conditioner Size and Performance, SAE Technical Paper Series #900219, February, 1990.
13. Akabane, H., et al., Evaluation of an Electircally Driven Automotive Air Conditioning System Using a Scroll Hermetic Compressor with a Brushless DC Motor, SAE Technical Paper Series #890308, February, 1989.

910251

A High Torque, High Efficiency CVT for Electric Vehicles

Frank A. Fitz and Paul B. Pires
Epilogics, Inc.

ABSTRACT

Epilogics, a young engineering firm in Los Gatos, CA, has developed the first fully geared, high torque, high efficiency, infinitely variable transmission suitable for automotive applications. The IVT has particular significance to electric vehicles because it can provide a highly efficient, yet exceptionally controllable means to regenerate power throughout the normal braking cycle (allowing regeneration even at near-zero vehicle speeds). Under normal operating conditions, the efficiency of the Epilogics transmission exceeds 90% as derived mathematically and corroborated experimentally. The device does not rely on traction to transmit torque and can therefore match the torque capacity of any typical gear drive. The size, weight, and cost of the device closely approximates that of a four-speed transmission suitable for the application. A recent development allows a bi-directional transmission which greatly enhances its attractiveness as an electric vehicle transmission by providing for a highly desirable (albeit, unnecessary in itself) means to regulate vehicle speed independently of motor speed, in addition to a means to a highly practicable regenerative braking system.

BACKGROUND

This paper presents the results of a study to determine the applicability of a continuously variable transmission to electricity-powered automobiles. A comparison was made between the computed deceleration and acceleration characteristics of a simplified vehicle modeled with a direct drive and with a CVT drive. The principle benefits of the CVT are a 58% improvement in vehicle energy recovery in the speed range of zero to 30 miles per hour (48 km/hr) and a 60% decrease in motor current during maximum possible acceleration.

VEHICLE DESCRIPTION

Since compact sedans represent the broadest range of interest, most current literature on electric vehicle development seems to focus on this class of vehicles. Therefore, typical compact sedan frontal area, weight and power were used for this study. The car was assumed to carry one or two passengers and only minimal cargo with a mix of highway and city street use. The performance needs were assumed to be modest -- fairly snappy acceleration off the line, dropping off rapidly above 30 miles per hour (48 km/hr). These considerations resulted in a passenger car as described in Table 1.

Table 1 - Vehicle Characteristics

Gross Weight	3,000	lb	1,360	Kg
Tire Outside Dia	20	in	0.5	mm
Frontal Area	20	ft^2	1.9	m^2
Drag Coefficient (Cd)	0.3			
Rolling Resistance	0.006			
Coefficient of Friction	0.6 (tires/pavemt)			
Acceleration				
0 to 30 mph (48 km/hr)	5	sec		
30 to 60 mph (96 km/hr)	11	sec		

DRIVELINE - Drivelines for the study were kept as simple as possible without sacrificing qualitatively valid results. In general, a single motor, conventional rear wheel drive layout is assumed with a motor just adequate to achieve the performance targets. The direct drive version layout is shown in Figure 1; the CVT version is identical except for the variable transmission shown in Figure 2.

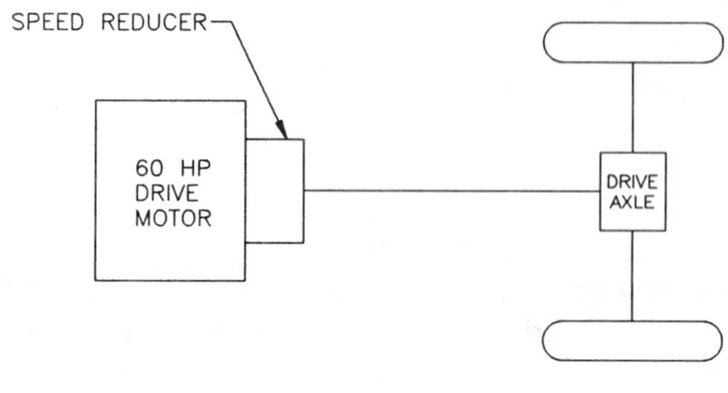

Figure 1

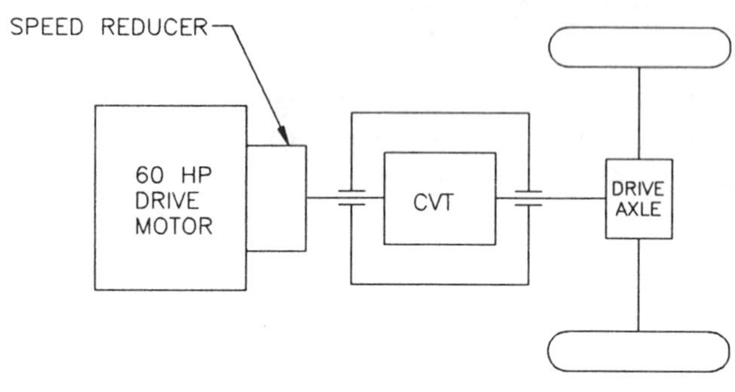

Figure 2

The detailed driveline assumptions used are listed in Table 2. The motor characteristics are for an idealized traction motor with the field connected in series. When driving, it is assumed that the speed voltage and torque current relationships are linear; no saturation effects are assumed within the range of interest. The behavior when operating as a generator during deceleration follows the simplified curve described in Table 2 and shown graphically in Figure 3 which is based on an assumption that

Table 2 - Driveline Characteristics

Motor Power	60	hp	45	Kw
Torque @ 6,000 rpm	52.5	ft-lb	71	N-m
Maximum Speed	7,000	rpm		
Back emf	24	volts/1,000 rpm		
Torque Constant	6.5	A/ft-lb	4.5	A/N-m
Efficiency	97	%		
Weight included in GVW				
Torque Table (In Regeneration Mode):				
0 to 1,000 rpm	0			
1,000 to 2,000 rpm		Linear Increase		
2,000 to 4,000 rpm	31.5	ft-lb	43	N-m
4,000 to 7,000 rpm	-4.5	% per 1,000 rpm		
Mechanical Components				
Motor Speed Reducer Ratio		2:1		
Rear Axle Ratio		3:1		
Efficiency		Negligible Loss		
Weight included in GVW				
CVT				
Ratio Range		1:1 to 1:4		
Efficiency	89	%		
Weight	90	lb	41	Kg

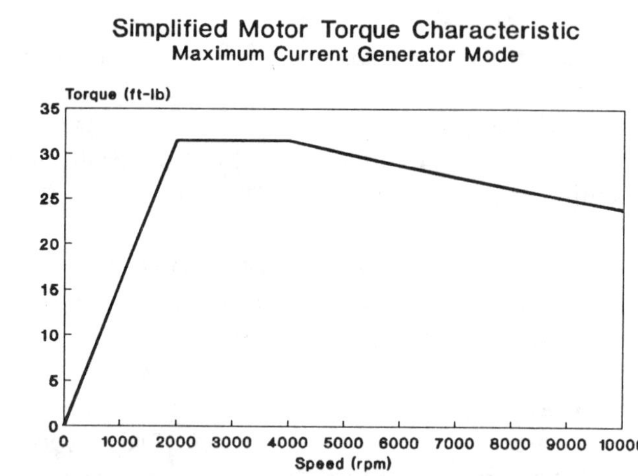

Figure 3

the motor is about 60% effective as a generator. The shape of the speed-torque characteristic curve is typical of a high capacity commercial alternator with an externally excited field.

The motor connects to the axle through a 2:1 speed reducer in both the direct drive version and the CVT drive version. For both cases, a 3:1 reduction drive axle routes power to the wheels. Losses in these two speed reducers are assumed to be negligible for the purposes of this study.

CVT - The CVT modeled is a fully geared,

Epilogics-type (1) transmission with a continuously variable ratio range of 1:1 to 1:4. The system is configured to provide up to a 4X motor speed increase when the wheels are driving the motor during deceleration. The ratio adjustment for this transmission can be controlled by a simple, servo-driven, linear actuator in concert with an electronic vehicle management system. Thus, any desired ratio can be readily available for any given operating condition.

Epilogics' CVT is a fully geared ratcheting-type transmission with the following characteristics:

1. Ratio adjustment varies linearly with control actuator displacement,
2. Actuation force is near zero; so ratio changes, even under load, can be readily effected with a simple servo system,
3. The transmission is hard geared throughout with no unique or unusual components required, and
4. Efficiency is comparable to any modern stepped-ratio transmission because of the positive gearing and conventional components.

CALCULATED PERFORMANCE

Regenerating performance and accelerating performance were determined over the speed range 0 to 60 mph (96 km/hr). Calculations were performed at 5 mph (8 km/hr) increments over this range. Where required, differentiation and integration were done by simple finite differences on the granularity of the calculation increments.

REGENERATION - Energy recovery potential was assessed by comparing maximum deceleration for the direct drive and the CVT drivelines. For each case, computations included the parameters listed in Table 3.

Table 3 - Calculated Parameters for Deceleration

Wheel Speed
Drive Shaft Speed
Motor Speed
Motor Torque
Regeneration Power
Speed Related Drag
Total Retarding Force
Deceleration
Elapsed Time
Regeneration Energy for Each Speed

The differences in results between the direct drive and the CVT are directly attributable to the capability to more efficiently manage the relationship of motor speed to vehicle speed with the CVT. The functional relationship used for this study is shown in Figure 4; the transmission ratio is controlled to keep the motor speed as close as possible to the maximum speed of 7,000 rpm throughout the vehicle speed range.

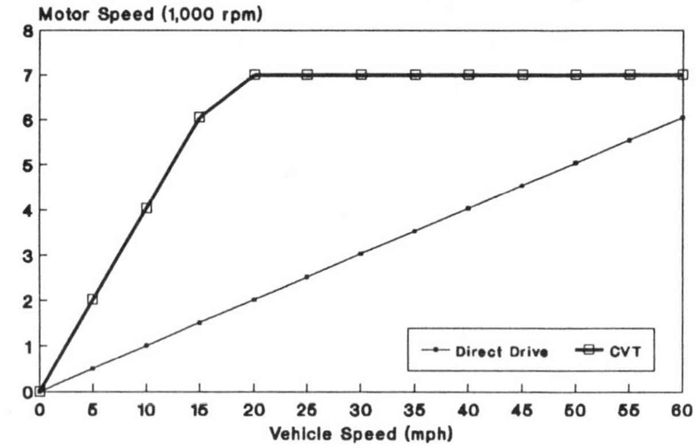

Figure 4

Figure 5 shows resulting vehicle deceleration as a function of vehicle speed. With direct drive, the maximum available deceleration of 3.5 ft/sec^2 (1 m/sec^2) occurs at the highest vehicle speed, having increased nearly linearly from 2.75 ft/sec^2 (0.84 m/sec^2) at 20 mph (32 km/hr). Below 20 mph, deceleration falls rapidly to almost zero. With the CVT, the highest deceleration of 8.6 ft/sec^2 (2.6 m/sec^2) occurs in the speed range

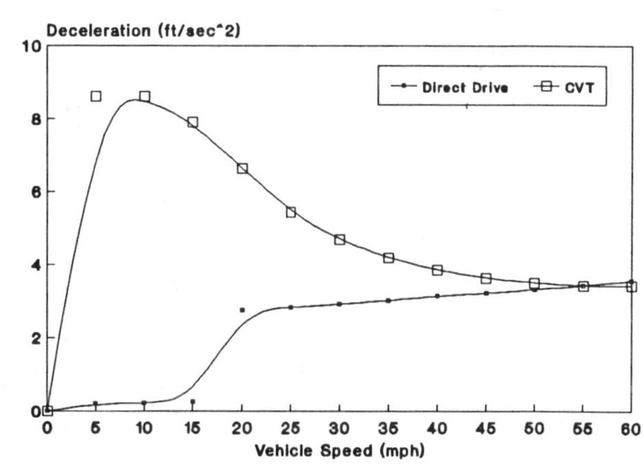

Figure 5

5 to 10 mph (8 to 16 km/hr), decreasing gradually at higher speeds to about the same value as the direct drive at 60 mph (96 km/hr). For purposes of the study, the CVT ratio control algorithm dictates a ratio of 1 to 1 at 60 mph; this results in the deceleration match with the direct drive at high speed.

Available regeneration power over the speed range is shown in Figure 6. Peak power of 14.5 Kw occurs at highest vehicle speed in both cases. Direct drive produces no significant regeneration power at speeds below 15 mph (24 km/hr). The CVT allows maximum power down to 15 mph and reduced but significant regeneration down to almost a standstill.

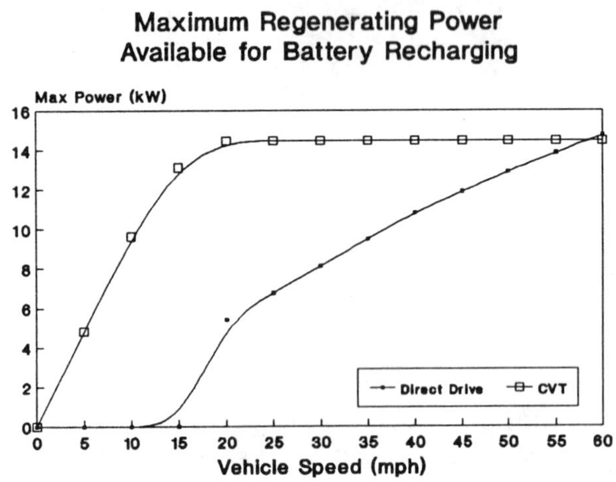

Figure 6

Regenerative energy available, speed increment by speed increment is shown in Figure 7. Recoverable energy for both the direct drive

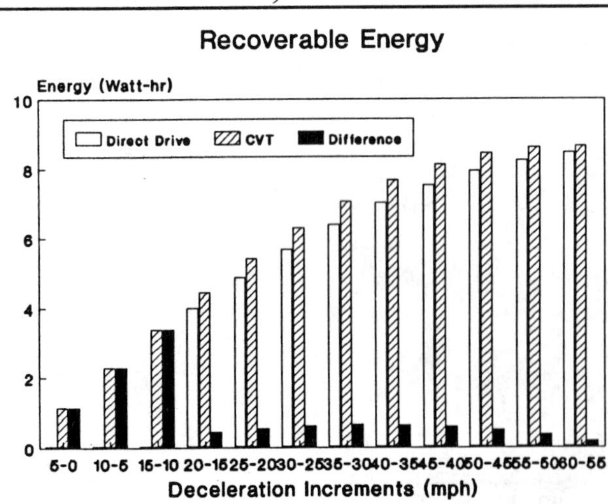

Figure 7

and the CVT is 8.5 Watt-hr for the 60 to 55 mph (96 to 88 km/hr) speed decrease. Intermediate speed ranges show decreasing but similar energy availabilities with a 5 to 10% difference in favor of the CVT. Below 20 mph (32 km/hr) the direct drive can provide no recoverable energy while the CVT provides an average of about 2 Watt-hr in the low speed range. A comparison based on a 0 to 30 mph (48 km/hr) city driving speed range indicates a 58% increase in recoverable energy when a CVT is used.

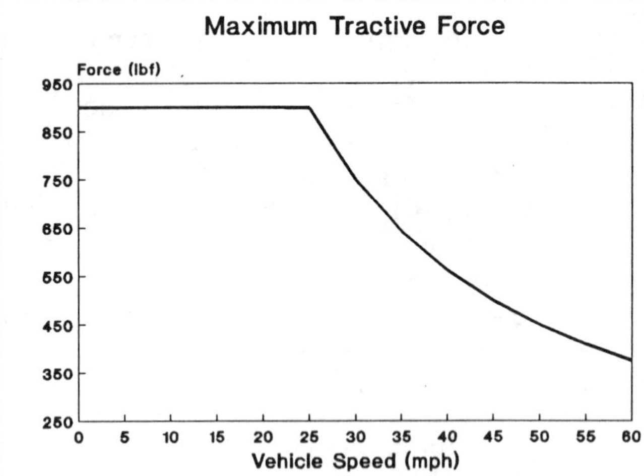

Figure 8

ACCELERATION - Potential performance in acceleration turns out to be identical for direct drive and the CVT due largely to the idealized characteristics chosen for the study motor and the assumption that the motor controller has no operational limitations. Acceleration is calculated using the maximum tractive force shown in Figure 8. A steady force of 900 lbf (4,000 N) from 0 to 25 mph (40 km/hr) is a result of tire to

Table 4 - Calculated Parameters for Acceleration

Drag Component of Tractive Force
Accelerative Component of Tractive Force
Total Tractive Force
Wheel Torque
Acceleration
Direct Drive Motor Torque
Direct Drive Motor Current
CVT Ratio Control Position
CVT Motor Speed
CVT Motor Torque
CVT Motor Current

pavement coefficient of friction assumption. Above 25 mph, the tractive force is motor power limited. Table 4 lists the computed parameters for this portion of the study.

Even though the performance is identical, motor speed is significantly different. With direct drive, motor speed is proportional to wheel speed. With the CVT, the ratio is controlled to maximum speed reduction until maximum motor speed was attained. Thereafter, ratio is controlled to maintain maximum motor speed. As a result, motor current is much lower with the CVT as shown in Figure 9. The maximum current (318 A) for the CVT drive is equal to the minimum current for the direct drive. Maximum current with the CVT is 42% of the maximum direct drive current.

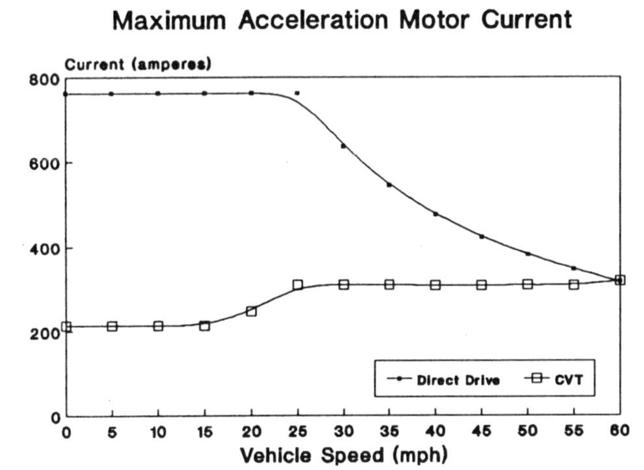

Figure 9

CVT EFFICIENCY - An efficiency of 90% is reasonable for this CVT. However, the effect of efficiency on the analysis (particularly the sensitivity of the results to changes in efficiency) required some analysis. An arbitrary value of 80% was selected for a sensitivity study. The essential finding is the effect on recoverable energy; this is shown in Figure 10 for comparison to the original 89% values shown in Figure 5. In the deceleration range 30 mph (48 km/hr) the energy recovery advantage over direct drive is 57% compared with 58% in the original analysis. At higher speeds, the CVT advantage erodes as speed increases and a slight disadvantage is seen at the maximum study speed. Peak available battery charging current drops to 90 amperes.

Additional losses from the low efficiency seriously impact the driveline's power demand under acceleration. Peak motor current increases

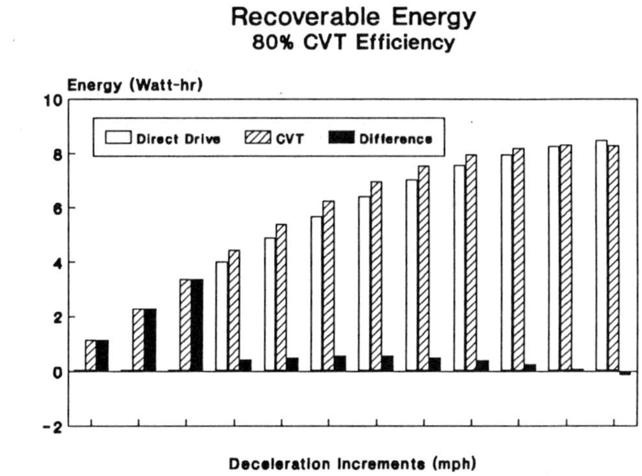

Figure 10

from the original 318 amperes to 360 amperes while sustaining the study acceleration performance target.

CONCLUSIONS

There is a high likelihood that use of a CVT with an overall efficiency greater than about 80% to enhance regeneration would significantly improve the range between charging of vehicles used primarily at low speeds in typical stop and go city street applications. Improvements at freeway and highway speeds appear to be too low to be of interest. Whether or not the apparent improvement can actually be achieved is related to the maximum charging current capability of the batteries in a vehicle. To the extent that this study resembles any actual vehicle, the batteries would need to be able to accept charging current of up to 100 A for short periods of time (a few seconds) with a charge duty cycle dependent on the target driving cycle. Energy storage devices like flywheels or gas springs which would be required with low charging current batteries were not considered.

The system benefits of using the CVT while accelerating and maintaining speed are less obvious although the high efficiency (about 90%) offers some advantage. The much lower motor current required should allow a reduction in complexity and weight of the final stages of the motor control electronics. At any fixed performance level, the lower required drive current also suggests the possibility of substantial motor size and weight savings. These controller and motor gains may further offset the added cost and system complexity introduced by a CVT.

REFERENCES

1. Fitz, Frank; Pires, Paul, "A Geared Infinitely Variable Transmission for Automotive Applications" SAE Paper presented at the SAE International Congress & Exposition, Detroit Michigan, February 1991.